THE
SMART GRID
FOR WATER

TREVOR HILL AND GRAHAM SYMMONDS

THE
SMART GRID
FOR WATER

HOW DATA WILL SAVE OUR WATER
AND YOUR UTILITY

Published by Advantage, Charleston, South Carolina.
Member of Advantage Media Group.

ADVANTAGE is a registered trademark and the Advantage colophon is a trademark of Advantage Media Group, Inc.

Printed in the United States of America.

ISBN: 978-159932-389-3
LCCN: 2013933412

This publication is designed to provide accurate and authoritative information in regard to the subject matter covered. It is sold with the understanding that the publisher is not engaged in rendering legal, accounting, or other professional services. If legal advice or other expert assistance is required, the services of a competent professional person should be sought.

Advantage Media Group is proud to be a part of the Tree Neutral® program. Tree Neutral offsets the number of trees consumed in the production and printing of this book by taking proactive steps such as planting trees in direct proportion to the number of trees used to print books. To learn more about Tree Neutral, please visit **www.treeneutral.com**. To learn more about Advantage's commitment to being a responsible steward of the environment, please visit **www.advantagefamily.com/green**

Advantage Media Group is a publisher of business, self-improvement, and professional development books and online learning. We help entrepreneurs, business leaders, and professionals share their Stories, Passion, and Knowledge to help others Learn & Grow™. Do you have a manuscript or book idea that you would like us to consider for publishing? Please visit **advantagefamily.com** or call **1.866.775.1696.**

FOREWORD

THE SMART GRID FOR WATER

How Data will Save Our Water and Your Utility

The water utility business is changing rapidly. Once described with adjectives such as stable, reserved, and stalwart, water utilities are experiencing a rapid change as they move into the twenty-first century. Faced with the challenges of managing water resource scarcity in the face of growth and climate variability, managing financial survival, and responding to the information demands of technology-savvy customers, utilities are discovering the need for, and power of, integrated data systems. Water utilities in the future will use faster access to better information to make better decisions and rapidly disseminate that information to customers and other stakeholders.

These changes are being driven largely by increasingly limited water availability and the high costs of developing alternative supplies. Water utilities are now charged with blazing a path to sustainability through efficiencies for the utility and the customer. As Dr. Peter Gleick of the Pacific Institute recently noted, the key to improving water efficiency is "understanding where, when, and why we use water."[1] Unfortunately, data of this granularity has been hard

---

1   Peter H. Gleick, "Roadmap for Sustainable Water Resources in Southwestern North America," PNAS 107, no. 50 (December 14, 2010): 21300–21305, www.pnas.org/cgi/doi/10.1073/pnas.1005473107

to come by in the water industry. In most utilities, data is a sparse commodity or is collected in such disparate platforms that interrelationships are missed or remain hidden.

The development of the smart grid for water is changing this paradigm by increasing the availability of data and information. The smart grid for water can tell us not only *how much* water is used, but *where* and *when*. The smart grid for water represents a significant shift from a data-poor, hardware-centric, asset-driven nineteenth-century model to a data-rich, information-centric environment that is 100 percent accurate and contextualized in space and time. This new approach has created the geotemporal data model, providing the where, when, and how necessary to understand water use, understand where opportunities for efficiencies exist, and engage the customer in conservation.

Utilities around the world are therefore grappling with a fundamental shift in operational philosophy. A change from asset-centricity to meter- or customer-centricity. Not only must utility operators consider the customer impacts of the rising costs of water on their customers, but in fact provide those customers with the tools to understand and manage their own behavior. Regulators around the world are incentivizing water utilities to further engage their customers in these ways while demanding customer conservation and better operating efficiencies from the utilities themselves.

It is for these reasons and our certainty of what the future of water management holds that we developed FATHOM™—the world's first geo-spatial, cloud-based, customer information system for water utilities. This fully integrated operating platform for water utilities draws on real-time data from all water utility operating data bases—volumetric, customer, financial and geo-spatial. This unique combination of real-time data allows utilities to lever and surface the value of Advanced Metering Infrastructure (AMI) data but also measure it in the context of time or dollars. This break-through in technology was born from our own operations and has had the benefit of years of development and testing in our own water utilities—systems that we have owned and grown for the past decade. It is this detailed operational knowledge, coupled with an often humbling journey of deploying advanced technologies in our own utilities that is the genesis of FATHOM. As a utility ourselves, we believe that bringing technology leverage and economies of scale to the highly fragmented water sector has the potential and promise to assist other utilities in achieving their goals of improving revenue, decreasing costs and increasing the levels of customer service within their own operations. FATHOM also provides a proven methodology for utilities to realize on their ambitions rapidly and economically.

FATHOM is the smart grid for water and I hope that our experiences offered in this book will provide a framework and an understanding of what the smart grid for water is, its potential and the means to support the business case to help drive other utilities into the twenty-first century.

For water utilities, focusing on collecting, aggregating, and analyzing data and converting it into real-time information will be critical. The twenty-first century water manager needs to manage the flow of data and information as well as the flow of water. Our future depends on it.

—TREVOR T. HILL

ACKNOWLEDGMENT

This book would not have been possible without the dedication and desire for improvement of all our staff at Global Water. Combining excellence in water management, data systems, customer service, engineering, compliance and operations, our team has together built, and in many cases rebuilt, our utilities into models for the twenty-first century.

CONTENTS

LIST OF FIGURES

P R E F A C E

The world today faces the enormous, dual challenges of renewing its decaying water infrastructure and building new water infrastructure. Now is an opportune moment to update the analytic strategies used for planning such grand investments under an uncertain and changing climate.[2]

—P. C. D. MILLY

Water, in its pure form, is a simple compound of two hydrogen atoms and one oxygen atom. This molecule, more than any other, is responsible for us. Its unique properties define our chemical processes, Earth's thermal characteristics, our engineering, and our lives. For most of humankind's tenure on this planet, water has defined where and how we live. Absent this molecule, we would not be having this discussion today.

The humble water molecule is the result of direct particle interactions from the big bang and stellar nucleosynthesis. Three-hundred and eighty thousand years after the big bang, the universe had cooled enough for simple hydrogen and helium atoms to form. Four hundred million years later, gravitational collapse drew these atoms together and fusion ignited the first stars.

2 P. C. D. Milly et al., "Stationarity Is Dead: Whither Water Management?" *Science 319* (2008)

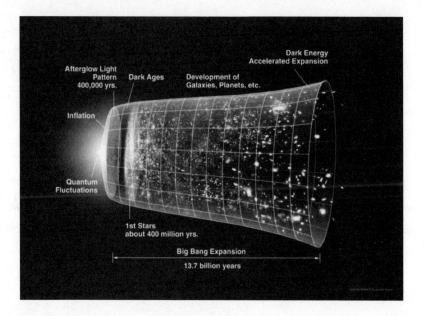

Figure 1: The big bang created the conditions for the formation of hydrogen and helium.[3]

Within those stars, hydrogen and helium were synthesized into neon, carbon, oxygen, iron, and other heavier elements. Stars shed these elements in nova and supernova, dispersing the heavier elements to the universe. Hydrogen and oxygen combined in energetic nebula clouds where soaring temperatures set off chemical reactions which resulted in the production of water molecules.

3   NASA/WMAP Science Team, http://www.jwst.nasa.gov/images/timeline_m.jpg

Figure 2: The production of water in space.[4]

In primordial solar systems, water molecules coalesced on dust and other particles and ultimately were delivered to planetary systems such as Earth through accretion or by collision with interplanetary bodies such as comets.[5] This process continues today throughout the universe. In fact, "water is the main constituent of the mantles on grains found in the low-density medium that fills the space between stars."[6]

---

4   European Space Agency. Cited in Elizabeth L. Chalecki, "Water and Space," in "The World's Water 2002–2003," biennial report, Pacific Institute. Courtesy ESA.

5   Henry H. Hsieh, "A Frosty Finding," *Nature 464* (April 29, 2010): *"Earth is thought to have formed dry owing to its location inside the 'snow line', which is the distance from the Sun within which it was too warm for water vapour in the nascent Solar System to condense as ice and be swept up into forming planetessimals. Therefore, the water that now fills our oceans and makes life itself possible must have been delivered to Earth from outside the snow line, perhaps by impacting asteroids and comets. The abundance of the hydrogen isotope deuterium in hydrated minerals found in certain meteorites is similar to the deuterium content of ocean water, suggesting that outer mainbelt asteroids are the most likely source of that water."*

6   Rachel Akeson, "Watery Disks," *Science 334* (October 21, 2011)

Despite its interstellar abundance, water remains a finite resource here on Earth. And while there may be a possibility for water delivery from an errant comet, that delivery comes with its own collateral issues. Just ask the dinosaurs.[7]

Figure 3: New sources of water: special delivery via comet or asteroid[8]

---

7   W. Alvarez, T. Rex and the Crater of Doom, Princeton University Press, 1997

8   http://www.donaldedavis.com/BIGPUB/BIGIMPCT.jpg (public domain) "NASA/Don Davis"

WHY WE NEED
SMARTER WATER

Three hundred and sixty three million trillion gallons (1.386 billion cubic kilometers) of water can be found within Earth's hydrologic cycle.[9] While that is a seemingly infinite reservoir of water, only 0.7 percent of this water is available for potential human access.[10] Throughout history, human civilization has been bounded by the spatial and temporal availability of water. The Romans, Mayans, Egyptians, and a host of other civilizations became very adept at moving water to where it was needed. Finally, in the industrial revolution and with the availability of cheap power, humans decoupled our spatial and temporal relationship with water. Today, however, intense urbanization (Asian cities alone are expected to grow by 1 billion people in the next twenty years), population growth (in the last century, world population has tripled, and is expected to increase from 6.5 to 8.9 billion by 2050), and the increasing volatility of water resources have put our engineered systems at risk. Indeed the spatial and temporal nature of water is again becoming increasingly important for humankind.

---

9   P. Gleick, "The World's Water 2000–2001," biennial report, Pacific Institute (2000)

10   http://www.climate.org/topics/water.html, accessed April 27, 2012

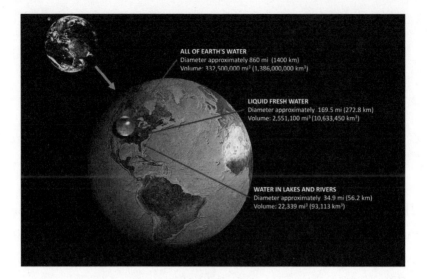

Figure 4: Availability of water on earth[11]

Figure 5: Drought in Texas[12]

---

11   Howard Perlman (USGS), based on globe illustration by Jack Cook (Woods Hole Ocean-ographic Institution), http://ga.water.usgs.gov/edu/2010/gallery/global-water-volume.html

12   ©Ann Worthy, 2013. Used under license from Shutterstock.com.

Compounding this issue is the fact that many utilities are caught between the need to invest in replacing aging infrastructure while at the same time suffering from the effects of the worst financial downturn since the Great Depression. Ironically, investing in infrastructure is critical to save water and money, but finding the capital dollars necessary is increasingly difficult. And the problem is only getting worse:

- The American Society of Civil Engineers estimates that by 2020, the capital infrastructure funding gap for water and wastewater will be $84 billion (and $144 billion by 2040).[13]

- The United States Geological Survey has reported that US water systems experience 240,000 water main breaks annually,[14] resulting in the loss of 1.7 trillion gallons of water every year. The US Environmental Protection Agency (USEPA) has estimated the cost of non-revenue water (NRW) to be $2.6 billion per year.[15]

- For developed countries, non-revenue water often represents 20 percent of the total water withdrawn from the environment. In developing nations, non-revenue water can account for as much as 50 percent due to distribution system leaks, theft, and poor measurement techniques.[16] The World Bank estimates that "the total cost to water utilities caused by NRW worldwide can be conservatively

---

13   American Society of Civil Engineers, "Failure to Act: The Economic Impact of Current Investment Trends in Water and Wastewater Treatment Infrastructure" (2011), http://www.asce.org/uploadedFiles/Infrastructure/Failure_to_Act/ASCE%20WATER%20REPORT%20FINAL.pdf

14   USEPA, "Addressing the Challenge Through Innovation," Office of Research and Development National Risk Management Research Laboratory

15   http://www.epa.gov/awi/distributionsys.html

16   http://www.pikeresearch.com/research/smart-water-meters

estimated at $141 billion per year" and represents enough water (45 million cubic meters per day) to serve nearly 200 million people.[17]

Notably, resource and financial pressures can tip rapidly, leaving utilities scrambling to address emergency water conditions or suffering intense financial hardship. Unfortunately, most utility solutions are engineered and are not "rapid response systems." They cannot be deployed rapidly and effectively to sustain water resources in the face of dramatically reduced runoff, or changes in snowmelt timing, or increased population, or financial crises. The velocity of these destabilizing influences far exceeds the ability of traditional solutions to react and compensate. In addition, as the degree of interconnectedness of systems and networks increase (in this case, the interconnection of resource and financial networks), they become in effect *less* stable:

> Surprisingly, a broader degree distribution increases the vulnerability of interdependent networks to random failure, which is opposite to how a single network behaves.[18]

To operate under this new paradigm, utilities must develop a set of practices and systems that are flexible, adaptable, and rapidly deployable in the face of the potential failures. In effect we need to bring a level of intelligence into our water systems that helps us identify problems, allows for robust mitigation, and gives us the information necessary to act, quickly.

---

17   Bill Kingdom, Roland Liemberger, Philippe Marin, "The Challenge of Reducing Non-Revenue Water (NRW) in Developing Countries. How the Private Sector Can Help: A Look at Performance-Based Service Contracting," World Bank, Water Supply and Sanitation Sector Board Discussion Series, paper no. 8, December 2006

18   S. Buldyrev, "Catastrophic Cascade of Failures in Interdependent Networks," *Nature* 464 (April 15, 2010): 1025–1028 , doi:10.1038/nature08932

RESOURCE AND FINANCIAL SCARCITY

SUPPLY-SIDE WATER MANAGEMENT

The developed world has been forged on the supply-side. In the past, we've assumed that we could engineer our way out of trouble. However, "when we build big, we build wrong."[19] This is because we typically build within a narrow band of expected conditions. Those assumptions, spurred by the relative passivity of the Holocene era, are being proven to be invalid for the future.[20]

For instance, in the United States, peak spring runoff from the snowshed of the Colorado River basin occurs an average of three weeks earlier than historic averages due to the recent five-fold increase in dust events on the snowpack. This earlier snowmelt allows for additional snow-free conditions that increase transpiration of the water,[21] and also strains the engineered collection systems which are designed to see a slower release of the stored water over a longer period.

---

19   Fred Pearce, *The Guardian,* February 24, 2012

20   P. C. D. Milly et al., "Stationarity Is Dead: Whither Water Management?" *Science 319* (2008)

21   National Science Foundation, "Windborne Desert Dust Falls on High Peaks, Dampens Colorado River Runoff," *ScienceDaily,* September 21, 2010, http://www.sciencedaily.com/releases/2010/09/100920172746.htm, accessed September 22, 2010.

Globally, changes in atmospheric temperature also drive an increase in the volatility of the water cycle, which is represented by intensification of weather events:

> The faster water cycles, the more abundant and more violent …storms might be. And wet places getting wetter can lead to more severe and more frequent flooding. Dry places getting drier would mean longer and more intense droughts.[22]

Recent research suggests that a 2° to 3° C increase in atmospheric temperatures will result in a 16 to 24 percent amplification of the global water cycle.[23]

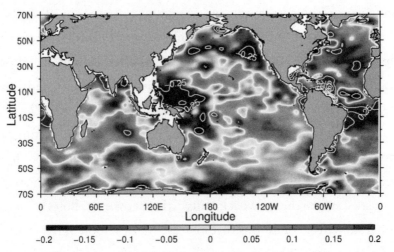

Figure 6: Changes in ocean salinity as an indicator of volatility of the water cycle[24]

Trying to adapt to this increasing volatility through engineering alone is difficult. Large-scale damming activities, pipelines from water-rich areas to water-poor areas, floating bags of water down coastlines,

---

22   R. Kerr, "The Greenhouse Is Making the Water-Poor Even Poorer," *Science 336* (April 27, 2012)

23   P. J. Durack et al., "Ocean Salinities Reveal Strong Global Water Cycle Intensification during 1950 to 2000," *Science 336* (2012): 455

24   Ibid.

pumping water across the continental divide, or hauling icebergs from the Arctic and Antarctic are not going to be economically or environmentally feasible.

A recent National Resources Defense Council (NRDC) report notes that there are inherent dangers in relying on the supply-side solution approach, exposing "serious questions about the reliability of surface and groundwater sources for proposed pipeline projects, including potential environmental impacts, existing constraints on water sources, and the likely impacts of climate change on these supplies."[25] The report concludes:

- Many of these projects rely on water sources that are far less reliable than past water projects.
- Some of these projects have the potential to increase conflict and harm other existing water users.
- By increasing reliance on unsustainable water sources, some of these projects could increase the water supply and economic vulnerability of communities in the long term.

The same arguments can be made with respect to desalination. While the source for desalinated water is vast, serious economic and environmental issues remain. Desalination is extremely power intensive: an acre-foot of desalinated water (325,851 gallons) consumes 4,000 to 9,000 kWh of energy.[26] In Saudi Arabia, desalination plants consume the equivalent of 1.5 million barrels of oil per day to generate the power necessary to operate the production

---

25   D. Fort, B. Nelson, "Pipe Dreams: Water Supply Pipeline Projects in the West," Natural Resources Defense Council, June 2012, http://www.nrdc.org/water/management/pipelines-project.asp

26   R. C. Wilkinson, University of California at Santa Barbara, *Southwest Hydrology*, September/October 2007

facilities.[27] That's 730 gallons of fuel per second. And in the perverse reality of the water/energy nexus, each barrel of oil requires 1,850 gallons of water to refine.

More importantly these supply-side projects take years—even decades—to complete, at tremendous environmental and fiscal cost, even when they are not in use. In Melbourne, Australia, despite catchments being full and the state government not ordering any water from a recently completed desalination facility, the costs are a staggering A$1.8 million per day.[28]

FINANCIAL SCARCITY

The water scarcity crisis is unfolding at a time of intense financial hardship for our municipalities. The municipal sector in the US has experienced a fiscal shortfall of between $56 billion and $83 billion from 2010 to 2012, driven by declining tax revenues, ongoing service demands and cuts in state revenues.[29] Compounding the current financial crisis for municipalities is the imperative of investment requirements for our aging infrastructure. Since 2002, it has been recognized that municipalities are faced with the challenge of broad-scale infrastructure replacement at a cost of $300 billion to $1 trillion dollars[30] over the next 20 years.

Protecting revenue and controlling costs means that we must fundamentally change our passive management of water and utility

---

27 http://www.circleofblue.org/waternews/2010/world/saudi-arabia-to-use-solar-energy-for-desalination-plants/#more-11102

28 http://www.bloomberg.com/news/2013-02-05/recycled-stormwater-among-australia-options-as-water-plant-idled.html

29 Christopher W. Hoene, "City Budget Shortfalls and Responses: Projections for 2010–2012," Center for Research and Innovation, National League of Cities, December 2009

30 General Accounting Office (GAO), "Water Infrastructure: Information on Financing, Capital Planning, and Privatization," report no. GAO-02-764, August 2002

resources and become active in ensuring we are efficiently monetizing the water cycle.

A COMBINED RAPID RESPONSE SYSTEM FOR RESOURCE AND FINANCIAL DROUGHT

Solving the water and financial volatility issues rapidly and effectively will be a hallmark of sustainable cities and provide a unique competitive advantage for those cities over other jurisdictions. Both issues can be mitigated through improving the temporal and spatial quality of water data, the tools of the smart grid for water:

- Fixing the data voids in our systems increases revenue.
- Actively monitoring the health of the meter population and proactively replacing meters preserves the accuracy of water consumption and protects revenue.
- Providing customers with highly granular, instantaneous data increases awareness of water use and decreases consumption.
- Reducing consumption extends the operational life of our water infrastructure and reduces the overall cost of infrastructure projects.
- Combining highly granular consumption data with customer information systems and georeferenced spatial data identifies instantaneous water loss.
- Increased data granularity allows physical and logical disconnects to be identified and corrected automatically, preserving the meter inventory integrity and protecting revenue.

THE CERTAINTY OF
UNCERTAINTY

A change to freshwater availability in response to climate change poses a more important risk to human societies and ecosystems than warming alone. Changes to the global water cycle and the corresponding redistribution of rainfall will affect food availability, stability, access, and utilization.[31]

—P. J. DURACK

WATER AVAILABILITY

Water has always defined the human landscape. Since the dawn of humanity, the presence of water has defined the presence of humans. The massive engineering works associated with civilizations such as the Egyptians, the Romans, the Hohokam, and the Maya inspired our own nineteenth- and twentieth-century responses to the age-old conundrum of getting water to the populace. The availability of water and ultimately the availability of power finally decoupled humans from the source of their water. This decoupling has actually re-exposed the water reality on Earth: while the amount of water in the Earth system is constant, it varies in quality, location, access, delivery, and availability. In fact, the distribution of fresh water is

---

31   P. J. Durack et al., "Ocean Salinities Reveal Strong Global Water Cycle Intensification during 1950 to 2000," *Science 336* (2012): 455

extremely disproportionate, with eleven countries sharing approximately 60 percent of the world's fresh water resources,[32] and 450 million people in twenty-nine countries suffer from chronic water shortages.[33]

Figure 7: Phoenix Basin Hohokam irrigation systems[34]

---

32 P. H. Gleick, "The World's Water 2008–2009," Table 1: Total renewable freshwater supply by country (2006 update), Pacific Institute, 2009; Brazil (14.9%), Russia (8.2%), Canada (6.0%), USA (5.6%), Indonesia (5.2%), China (5.1%), Colombia (3.9%), Peru (3.5%), India (3.5%), Congo (2.3%), Venezuela (2.2%)

33 World Economic Forum, "Global Risks at a Glance," http://www.weforum.org/pdf/CSI/Risks_at_a_Glance.pdf

34 Photo and caption from http://www.cdarc.org/what-we-do/archaeology-southwest/archaeology-southwest-vol-21-no-4/ © Adriel Heisey

Figure 8: Hoover Dam construction[35]

By 2025 the United Nations Food and Agriculture Organization (FAO) estimates 1.8 billion of the world's projected 8.9 billion people will be living in countries or regions that are experiencing "absolute water scarcity," and two-thirds of the world population could be under conditions of water stress.[36]

In some areas the conditions are particularly dire: "Yemen has the second-fastest population growth of any country on Earth. It also has very little water. In fact, the country's capital, Sana'a, is expected to have exhausted its underground aquifers within just six years…so

35 US Bureau of Reclamation. Photo from http://www.ecommcode.com/hoover/hooveronline/hoover_dam/const/087.html

36 United Nations Food and Agriculture Organization (UNFAO), "Coping with Water Scarcity," March 22, 2007; http://www.fao.org/nr/water/docs/escarcity.pdf

Sana'a is expected to be the world's first capital city to completely run out of water."[37]

Number of months in
which water scarcity > 100%

- 0
- 1
- 2 - 3
- 4 - 5
- 6 - 7
- 8 - 9
- 10 - 11
- 12
- No data

Figure 9: World water stress[38]

Whereas water scarcity is a daily thought in developing countries, the developed world is not immune to water crises. Pike Research has indicated that "water scarcity is a looming issue that will affect nearly half the world's population by 2030. In the United States, the problem is even more near term, with 36 states expected to face water shortages by 2013."[39] Climate change and the variability of water sources are being felt from Australia to Atlanta. A recent Australia-Davos report notes "water scarcity may well pose the single biggest threat to required economic growth and future international competitiveness."[40] In the American Southwest, prolonged drought has driven Colorado River reservoirs to extreme levels as inflows to these reservoirs have decreased substantially. Lake Mead, the largest reservoir on the Colorado River system, is at its lowest level since

37 *The Telegraph*, "After food protests, water riots are next," January 31, 2011, http://www.telegraph.co.uk/finance/newsbysector/retailandconsumer/8291454/After-food-protests-water-riots-are-next.html

38 Hoekstra AY, Mekonnen MM, Chapagain AK, Mathews RE, Richter BD (2012) Global Monthly Water Scarcity: Blue Water Footprints versus Blue Water Availability. PLoS ONE 7(2): e32688. doi:10.1371/journal.pone.0032688

39 http://www.pikeresearch.com/research/smart-water-meters

40 Australian Davos Connection, "Australia Report: Risks and Opportunities," 2010, 3, http://www.ausdavos.org/files/Documents/ADC%20Australia%20Report%202010.pdf

1965 and has dropped 130 feet in the last decade.[41] While some reprieve has occurred due to a healthy snow pack, the reliability of the Colorado River as the lifeblood of the US Southwest must be questioned, particularly in the context of increasing climate volatility.

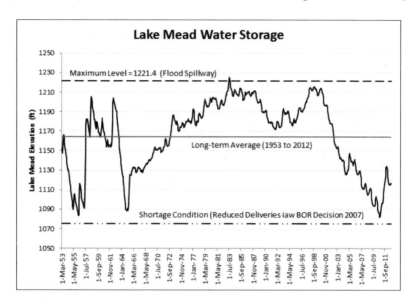

Figure 10: Lake Mead water levels[42]

Research indicates that long-term droughts are not unusual in this region, and in fact have occurred throughout earth's history[43]—most recently in what is known as the Medieval Climate Anomaly.[44] Flows in the Colorado River have been decreasing throughout the twentieth and twenty-first centuries, averaging in the order of fifteen

41 Brett Walton, "Low Water May Halt Hoover Dam's Power," Circle of Blue, September 22, 2010; http://www.circleofblue.org/waternews/2010/world/low-water-may-still-hoover-dam%E2%80%99s-power/

42 Lake Mead at Hoover Dam, Elevation (http://www.usbr.gov/lc/region/g4000/hourly/mead-elv.html)

43 Peter J. Fawcett et al., "Extended Megadroughts in the Southwestern United States during Pleistocene Interglacials," *Nature 470* (February 24, 2011), doi:10.1038/nature09839

44 D. M. Meko, C. A. Woodhouse, C. A. Baisan, T. Knight, J. J. Lukas, M. K. Hughes, and M. W. Salzer, "Medieval Drought in the Upper Colorado River Basin," *Geophysical Research Letters 34* (2007): L10705, doi:10.1029/2007GL029988

million acre-feet (MAF) annually,[45] far below the 22 MAF of annual flow extant during the time that the Colorado River was formally allocated to the seven basin states[46] by the Colorado River Compact in 1922[47] and to Mexico under the Mexican Water Treaty of 1944.[48] And far below the legally allocated volume of 16.5 MAF. Recent studies by the Scripps Institution of Oceanography have suggested that the sustainable inflow in the Colorado River is in the order of 11 to 13.5 MAF.[49]

But it is not only the physical scarcity of this precious resource that is causing concerns. The growing variability and volatility in the supply of renewable water is increasingly problematic: overloading engineered systems that were based on the expectation of stable, reliable weather and water delivery patterns.[50]

The United States Bureau of Reclamation (USBR) recently released an assessment of the climate change implications for several river basins including the Colorado, which states: "the Southwestern United States to Southern Rockies are projected to experience gradual runoff declines during the 21st century."[51]

The USBR report continues: "Based on current reservoir operational constraints (e.g., storage capacity, flood control rules, constraints on reservoir water releases to satisfy various obligations),

---

45   USBR Colorado River flow data

46   States of Arizona, California, Colorado, Nevada, New Mexico, Utah, and Wyoming

47   The Colorado River Compact of 1922 established that the upper basin and lower basin states would each be allocated 7.5 million acre-feet annually.

48   The Mexican Water Treaty allocated 1.5 million acre-feet of Colorado River water to Mexico.

49   Tim P. Barnett and David W. Pierce, "Sustainable Water Deliveries from the Colorado River in a Changing Climate," PNAS 106, no. 18 (May 5, 2009): 7334–7338

50   Brisbane, Australia, most recently in the grips of a massive drought that required the curtailment of all but essential water use, has since seen incredible deluges resulting in catastrophic flooding.

51   Reclamation, SECURE Water Act Section 9503(c): "Reclamation Climate Change and Water," Report to Congress, April 2011:viii

it appears that projected reductions in natural runoff and changes in runoff seasonality would lead to reduced water supplies under current system and operating conditions."[52]

The velocity and timing of runoff are altering the ability of our infrastructure to support our needs. Our storage infrastructure is based on historic runoff schedules and those are changing. "Warming trends appear to have led to a shift in cool season precipitation towards more rain and less snow, which has caused increased rainfall-runoff volume during the cool season accompanied by less snowpack accumulation in some Western United States locations."[53] Snowpack acts like a storage system for water; it slowly releases water in the spring just as demand is ramping up. Change the availability of the "snowpack storage" and the total available storage in the system as a whole decreases.

While the USBR report indicates that "water management systems across the West have been designed to operate within envelopes of hydrologic variability," it also notes that "the ability to use storage resources to control future hydrologic variability and changes in runoff seasonality is an important consideration in assessing potential water management impacts due to natural runoff changes."[54]

USBR is predicting a reduction in runoff from 3.1 to 8.5 percent. That doesn't sound like much, but when you combine it with the recent studies by the Scripps Institution of Oceanography that demonstrate "if climate change reduces runoff by 10%, scheduled deliveries [of Colorado River water] will be missed 58% of the time by

---

52 Reclamation, SECURE Water Act Section 9503(c): "Reclamation Climate Change and Water," Report to Congress, April 2011:36

53 Reclamation, SECURE Water Act Section 9503(c): "Reclamation Climate Change and Water," Report to Congress, April 2011:vii

54 Reclamation, SECURE Water Act Section 9503(c): "Reclamation Climate Change and Water," Report to Congress, April 2011:31

2050. If runoff reduces 20%, they will be missed 88% of the time,"[55] the real potential for major disruptions of Colorado River supply exists.

These analyses also do not take into account the projection that the increased dust impacts on snowmelt in the Colorado River basin will reduce runoff by five percent,[56] and the fact that the peak spring runoff from the Colorado River basin is occurring an average of three weeks earlier due to dust.[57]

While variability in surface water flows is relatively apparent, the impact on groundwater sources is much less so. Groundwater is harder to quantify. In many cases we are depleting fossil water at an unsustainable rate.

The Stockholm Environment Institute notes that "at today's rates of water use [2011], the Southwest is projected to use 1,303 million acre-feet of groundwater," resulting in an overdraft of 260 million acre-feet. What this means is that in the western United States, we are expecting to use 85 trillion gallons of groundwater more than what is considered sustainable. When considering the effect of growth and climate change on demand, the expected overdraft in the year 2100 may be as high as 2,253 million acre-feet.[58]

The impacts of this increased groundwater use go beyond depletion of Earth's fossil water reserves. A US Geological Survey

55   Tim P. Barnett and David W. Pierce, "Sustainable Water Deliveries from the Colorado River in a Changing Climate," PNAS 106, no. 18 (May 5, 2009): 7334–7338.

56   Thomas H. Painter, Jeffrey S. Deems, Jayne Belnap, Alan F. Hamlet, Christopher C. Landry, and Bradley Udall, "Response of Colorado River Runoff to Dust Radiative Forcing in Snow," PNAS (2010), 20 September 2010, doi: 10.1073/pnas.0913139107

57   National Science Foundation, "Windborne Desert Dust Falls on High Peaks, Dampens Colorado River Runoff," ScienceDaily September 21, 2010, http://www.sciencedaily.com / releases/2010/09/100920172746.htm, accessed September 22, 2010

58   Frank Ackerman, Elizabeth A. Stanton, "The Last Drop: Climate Change and the Southwest Water Crisis," Stockholm Environment Institute—US Center, 2011:5, http://www. sei-us.org/publications/id/371

(USGS) report on the availability of groundwater stated that over-drafting aquifers results in "increased pumping costs, deterioration of water quality, reductions of water in streams and lakes, and land subsidence."[59]

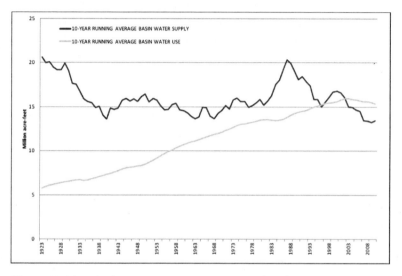

Figure 11: Historical 10-Year Running Average Colorado River Basin Water Supply and Use[60]

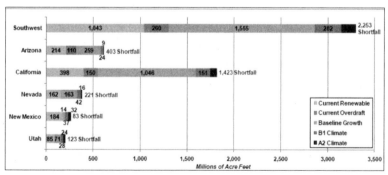

Figure 12: Southwest USA groundwater deficits[61]

---

59   T. E. Reilly, K. F. Dennehy, W. M. Alley, and W. L. Cunningham, "Ground-Water Availability in the United States," USGS Circular 1323 (2008): 70; also available online at http://pubs.usgs.gov/circ/1323/:15

60   US Department of the Interior, Bureau of Reclamation, "Colorado River Basin Water Supply and Demand Study - Study Report", December 2012

61   Frank Ackerman and Elizabeth A. Stanton, "The Last Drop: Climate Change and the Southwest Water Crisis," Stockholm Environment Institute—US Center, 2011:5, http://www.

Figure 13: Areas of water-table decline or artesian water-level decline in excess of 40 feet[62]

So combined we are experiencing:

- Decreased snowpack
- Increased temperature
- Increased dust
- Advanced snowmelt
- A shift to more winter water and less snow-derived water (reduction in the storage)
- Increased fossil groundwater use
- Decreasing groundwater quality

All this leads to increased variability and volatility in Western water supply.

---

sei-us.org/publications/id/371

62   T. E. Reilly, K. F. Dennehy, W. M. Alley, and W. L. Cunningham, "Ground-Water Availability in the United States," *US Geological Survey Circular 1323* (2008): 70; also available online at http://pubs.usgs.gov/circ/1323/, 15

But it is not just the Southwest that faces an uncertain water future. Water scarcity also threatens agriculture, growth, productivity, and business survivability even in areas that are considered "water rich." For example, in May 2008, Brockton, Massachusetts, a town that receives four feet of rain annually, commissioned a $60 million reverse osmosis plant to create potable water from seawater to attain a measure of reliability for their water source. [63]

GROWTH

Superimposed upon water scarcity are continued population growth, and concomitant changes in the fabric of the landscape: conversion of natural areas to urban, suburban, and commercial living spaces. The US population grows at approximately one percent[64] or approximately 3,000,000 people per year, which is occurring with a continued demographic shift to areas of lower water availability.

---

63   Circle of Blue, "U.S. Faces Era of Water Scarcity", July 9, 2008, http://www.circleofblue.org/waternews/2008/world/us-faces-era-of-water-scarcity/

64   Derived from US Census data file, http://www.census.gov/popest/national/files/NST_EST2009_ALLDATA.csv. Average growth rate from 2001 to 2009 was calculated as 0.97 percent.

Figure 14: Population growth and water availability (2000 to 2020)[65]

As a result, many jurisdictions are seeking means by which to become increasingly self-sufficient and live within available water supplies. This requires a change in the way that water managers view their resource. The combined effects of increasing demand and increasing volatility in supply mean that we must look beyond the historic paradigm of solving water issues through supply augmentation. We must look to curbing demand as an active, rather than reactive, water management strategy. Indeed it is more likely that we solve the problem of finite water availability with demand-side management than it is by building supply.

---

65   The Artemis Project, "Water Matters: Venture Investment Opportunities in Innovative Water Technology," research report, 2008. Originally published by M. Chan, US Department of Energy, National Energy Technology Laboratory, 15 March 2002.

WHY DO WE PAY FOR WATER?

The necessity to provide water for life is often viewed in opposition to its continuously rising cost. As ancient cities grew beyond their natural sources (rivers, lakes, and wells), water delivery and distribution became one of balancing the availability of water, the costs of providing that service, and the infrastructure necessary to get the water to the people. The water utility was born. And it was born to provide as much water to people and businesses as they could possibly use—without restriction.

In the thirteenth century, the City of London embarked on the construction of twelve "conduits" designed to bring water from springs outside the city to cistern houses. From these cisterns, the water was piped to larger storage facilities equipped with cocks or taps for dispensing the water. In the early fourteenth century, it was decided that brewers, cooks, and fishmongers should pay for the water they used, at the discretion of the keeper of the conduit, whose chief duty was to ensure that the water was not stolen for commercial purposes.[66] Water "cobs" hand-delivered water on a fee-for-service basis from the cisterns to the city's residents.

The growing population soon outstripped the existing conduits, and the polluted Thames River was increasingly less desirable as a source. The solution was to construct the New River canal system to

---

66   Roger D. Hansen, "Water-related Infrastructure in Medieval London," PDF:4–5, www. waterhistory.org

bring water from springs in the countryside to the city. This 60 km canal was enormously expensive, and required significant investment. The king provided half the funds (for which the royal household would receive "*for ever the one halfe of the benefitt profitt*"). A further twenty-nine investors completed the funding and the aqueduct and distribution system was completed in 1613. It was not until 1633 that the New River Company began distributing profits to its shareholders. The infrastructure costs far exceeded the carrying capacity of the revenue derived from the system.

These examples demonstrate from a very early point the financial complexity of water service provision: infrastructure is massively expensive; investors demand returns; and ensuring the cost recovery of operating and maintaining delivery systems is essential to sustaining service.

The same holds true today. Water remains the most capital-intensive utility business in which to operate. The National Association of Water Companies notes that "for a water utility to earn a dollar, nearly \$3.40 must be invested in infrastructure, an intensity that approaches an average of three times that of other utility sectors."[67]

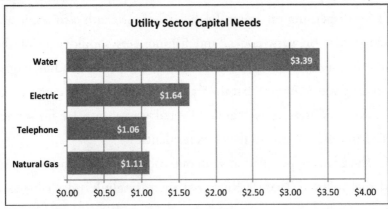

Figure 15: Capital needs by sector[68]

67   National Association of Water Companies, "Price, Cost, Value"
68   Ibid.

THE COST OF WATER

The Certainty of Higher Water Rates

Increasing demand, decreasing quality, increasingly stringent regulations, deteriorating infrastructure, and the future impact of climate induced variability are straining water resources, and the utility's financial capacity to address them. The cost of acquiring, producing, treating, and delivering water is going up. The result: higher prices for customers.

From a water resources perspective, this is not a bad thing. Without the right price signals, profligate water use will occur. We certainly know that if you make water cheap enough, people will use it—all of it, every last drop.

Dr. Gleick pointed this out in a 2010 article:

> Fresno's water rates are among the lowest, and their water use among the highest, of anyone's in California. Average Fresno residential use is 290 gallons per person per day. The state average is 135. For the same amount of water (22,440 gallons, more than enough for a family of four for a month) City of Fresno customers pay, on average, a monthly water rate of only $28.33, compared with San Francisco's $89.57 and San Diego's $95.48.[69]

---

69   Peter Gleick, "Smart Water Meters, Dumb Meters, no Meters," *SFGate*, April 28, 2010, http://www.sfgate.com/cgi-bin/blogs/gleick/detail?blogid=104&entry_id=62392

Making water a "financially valuable" product from a customer perspective while at the same time providing the necessary information to make appropriate consumption choices will be critical for resource sustainability.

COST OF TREATMENT

All potable water sources are required to meet minimum public health requirements. These regulatory requirements are becoming increasingly more stringent, while at the same time water sources are increasingly being degraded by human activity. The USEPA is required by statute[70] to maintain a Contaminant Candidate List (CCL) and evaluate a minimum of five contaminants on the CCL for the basis of a regulatory determination (based on the potential for human health impacts).

Recently, the USEPA has taken renewed interest in fluoride, perchlorate, nitrosodimethylamine (NDMA), pharmaceutically active compounds, hexavalent chromium, and other contaminants. Treatment to meet new or reduced maximum contaminant levels (MCLs) will require the introduction of specialized treatment systems and increase costs.

The fact that detection technology precision is constantly improving means that there will be a continuously increasing ability to detect constituents (down to parts-per-trillion, -quadrillion, and -quintillion levels). Finding these constituents in water will drive regulatory agencies to regulate them. Once regulated, the removal, sequestration, and disposal of these constituents represent a significant cost for utilities. These costs will ultimately be passed to the customer.

---

70   Safe Drinking Water Act

AGING INFRASTRUCTURE

Much of our water and wastewater infrastructure was installed generations ago. And much is reaching the end of its useful life. Municipalities are facing the challenge of broad-scale infrastructure replacement at a cost of $300 billion to $1 trillion dollars.[71] While we have managed to build substantial cities and municipalities on the backs of the infrastructure installed by our grandfathers, the days of reckoning are approaching. Our entire infrastructure—bridges, roads, pipes—is reaching end-of-life. Replacement of this infrastructure will not be cheap. The fact is that municipalities will be replacing infrastructure for decades, and resources, financial and human, will be consumed by the response to this issue.

This is not simply an issue of pipe replacement; it is a public health concern. Over the past thirty years, there has been marked reduction in the number of water-borne disease outbreaks attributed to water treatment systems. Over the same period, however, the percentage of incidents of disease outbreak as a result of distribution system deficiencies have increased exponentially.

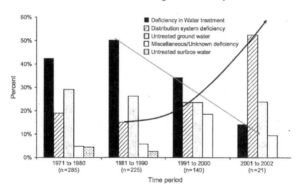

Figure 16: Sources of water-borne disease outbreaks shifting to distribution systems[72]

71   GAO, "Water Infrastructure: Information on Financing, Capital Planning, and Privatization," report no. GAO-02-764 (August 2002)

72   Reproduced from M. F. Craun et al. (2006), "Waterborne Outbreaks Reported in the United States," *Journal of Water and Health 4 (Suppl. 2)*: 19–30 with permissions from the

The reason for this can be traced directly to water-main leaks, failures, and losses of pressure. There is a very strong association of water-borne disease outbreaks with reporting of loss of pressure related to burst water mains.[73] The fact is that the conditions under which repairs are made are far less sanitary than those of new construction, resulting in a high probability of contamination from the surrounding materials. A study of the environment (water and soil samples) surrounding water mains noted that fecal coliform bacteria were detected in 43 percent of water samples and 50 percent of the soil samples. Fifty-six percent of these samples were also positive for viruses.[74]

As a result, we must view the investment in water infrastructure as connected to our entire urban existence. Not only is our ability to detect contaminants increasing, driving up the costs of regulatory compliance, our aging infrastructure is placing whole populations at a health risk. Neither of these issues can be ignored.

POWER COSTS

The power intensity of water is directly related to:

- The source
- The required treatment to bring it to potable standards
- The use

As the source water deviates from being ready for direct potable use, the power costs of production increase dramatically. For instance,

---

73   Paul R. Hunter et al., "Self-Reported Diarrhea in a Control Group: A Strong Association with Reporting of Low-Pressure Events in Tap Water," *Clinical Infectious Diseases* 40(4): e32–e34

74   Mohammad R. Karim, Morteza Abbaszadegan, and Mark LeChevallier, "Potential for Pathogen Intrusion during Pressure Transients," *Journal of American Water Works Association* 95:134–46

desalination certainly has its place in the world of water, but it is far from a panacea. Creating potable water from seawater can take as much as 4,000 to 9,000 kWh per acre-foot (12.3 to 27.6 kWh/1000 gallons).[75] Indeed many desalination facilities are constructed with their own power production plants to produce this power. With power costs continually on the rise, due to fuel cost increases, regulatory issues (e.g., air pollution regulations) and increasing demand, power for water treatment, and delivery will continue to increase the costs of water service for customers.

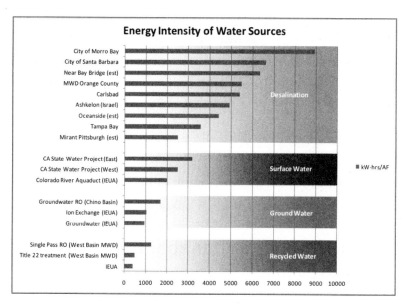

Figure 17: Energy intensity of water[76]

75   R. C. Wilkinson (University of California at Santa Barbara), "The Energy Intensity of Water Supplies" (sidebar in "California's Energy-Water Nexus"), *Southwest Hydrology 6, no. 5* (September/October 2007)

76   Data has been compiled based on sources published by: URS Corporation, California Coastal Commission, California Energy Commission, University of California — Santa Barbara, California Institute for Energy Efficiency, Marin Municipal Water District — Board of Directors, Bahman Sheikh, Ph.D, Robert C. Wilkinson, Ph.D, and Bruno Sauvet Goichon

TWENTIETH CENTURY RATES:
THE CERTAINTY OF DEMAND DESTRUCTION

While most municipalities have means for recovering costs of the provision of water service, the underlying rate structures, and the principles and goals vary widely. And if the water resources were unlimited, the most economically efficient rate structure could be determined with each customer class paying only for those needs specific to its use.

Unfortunately, water utilities are in the position of managing an increasingly scarce resource, and must now operate not on the basis of a regular business (i.e., increasing sales to increase revenue), but in a system that permanently destroys demand for the sake of resource sustainability. How to do this without also destroying the financial viability of the utility will be the fundamental problem of twenty-first century utility management, particularly since the infrastructure requirements for water sustainability (water reuse, dual water mains, control systems, treatment) and water suitability (treatment, etc.) will continue to increase the cost of providing service to customers.

BEHAVIOR MODIFICATION TOOL FOR ACHIEVING
FINANCIAL AND WATER RESOURCE SUSTAINABILITY

To be successful in the twenty-first century, utilities must encourage their customers to be aware of, and take stewardship of, their water use. Changing the behavior of the customer will be very important, particularly since we have, for the most part, isolated the general public from knowledge about water. Peter Gleick summed it up thus:

> We've done a great job in the United States in reducing our vulnerability to drought by building massive infrastructure for storing water in wet periods, so we can use it in dry periods.

The downside is it's made us, in a sense, ignore water as an issue for far too long. We now see growing water shortages not just in places that we used to think were dry, but in places that we used to think were wet. We're ignoring the way we use water. We've stopped thinking about how to use water efficiently and effectively because we've always assumed that it would always be there. That no longer is the case. We're moving into an era, truly, I think, of water scarcity throughout the United States and that by itself is going to force us to move to an era of more efficient management techniques.[77]

Changing behavior is a complex task. In a recent research article, Thomas Dietz described the complexity of behavioral change related to carbon reduction. The application of these lessons to water conservation is equally strong: to be successful, water managers will need to touch customers many times and in many ways:

Mass media appeals and informational programs can change attitudes and increase knowledge, but they normally fail to change behavior because they do not make the desired actions any easier or more financially attractive. Financial incentives alone typically fall far short of producing cost minimizing behavior...However, *interventions that combine appeals, information, financial incentives, informal social influences, and efforts to reduce the transaction costs of taking the desired actions have demonstrated synergistic effects beyond the additive effects of single policy tools* (emphasis added).[78]

---

77   Circle of Blue, "Present U.S. Water Usage Unsustainable: An Interview with Dr. Peter Gleick," July 8, 2008, http://www.circleofblue.org/waternews/2008/world/north-america/present-us-water-usage-unsustainable-an-interview-with-dr-peter-gleick/

78   Thomas Dietz, Gerald T. Gardner, Jonathan Gilligan, Paul C. Stern, and Michael P. Vandenbergh, "Household Actions Can Provide a Behavioral Wedge to Rapidly Reduce US Carbon Emissions," PNAS 106, no. 44 (November 3, 2009): 18452–18456, published online

In short, it is imperative to use the influence of education, information, and incentive packages to change behavior. And they have to be easy for the customer to implement and use.

While the majority of the reasons to conserve have historically been altruistic, there now exist powerful, tangible reasons for conservation, including real scarcity, growing populations, and a desire to reduce costs. Further, if a financial incentive can be developed such that customers are rewarded for conservation, the untapped power of positive reinforcement can be brought to bear on water use.

The smart grid for water offers the unique opportunity to combine both elements: knowledge of the financial impact of profligate water use; and the understanding of the social context of water use. Price signals and peer pressure, delivered on a regular basis, represent the only way to change customer behavior about water.

However, for utilities, in many cases, rate designs are such that achieving conservation via increased customer costs inevitably result in decreasing revenues for the utility. Establishing a rate structure that destroys demand without destroying the utility financially is critical for our water utilities to continue to manage this vital resource.

The conundrum for customers is that conservation does not automatically mean reduced costs for the end user. We must be able to communicate that water scarcity, treatment, and infrastructure will most certainly drive up costs, but it is behavioral pattern change that will achieve sustainability. Properly capitalized, financially solvent utilities are the cornerstone for the investment we need in sustainable water infrastructure and in the end protect the health of the public.

before print October 26, 2009, doi: 10.1073/pnas.0908738106, www.pnas.org/cgi/doi/10.1073/pnas.0908738106

PRICE ELASTICITY OF WATER

One of the challenges of developing conservation oriented rates is the fact that, at current prices, demand is unresponsive to price. However, Olmstead and Stavins note that any estimate of price elasticity "represents an elasticity in a specific range of prices."[79] That is, if the total cost is low, the response to increasing costs will be low. As prices increase—and they surely must—responsiveness increases. This presents some interesting rate design issues, as elasticity will begin to occur at higher tier levels within the same classes of customers.

In general, the long-term impacts of price elasticity are greater than the short-term elasticity. This is a result of the time required for customers to react to the implementation of higher rates, and affect the completion of the necessary modifications (behavioral and infrastructure) to achieve reductions in consumption. For example, if a large step-function increase in rates is applied, customers may immediately be able to reduce some discretionary consumption, resulting in a dip in demand. However, to complete the reaction, in some cases retrofitting fixtures, replacing landscaping, increasing internal water reuse will occur, and that process will take time.

On average, a 10 percent increase in the marginal cost of water can be expected to reduce residential demand by 3–4 percent in the short run. In the long term, such an increase could be expected to yield a 6 percent decrease in demand.[80] (Note that price elasticities are subject to a wide number of variables, including household population, household income, lot size, landscaping use, temperature, precipitation, and so forth, and may not be directly translatable between

79 Sheila M. Olmstead and Robert N. Stavins, "Comparing Price and Nonprice Approaches to Urban Water Conservation," *Water Resources Research 45*, *W04301* (April 25, 2009):4, doi:10.1029/2008WR007227

80 Ibid.

jurisdictions. Therefore "price elasticities must be interpreted in the context in which they have been derived."[81])

Clearly, price sensitivity to water resulting in demand reductions will reduce utility revenue. A true conservation oriented rate structure must take into account this revenue destruction that is concomitant with demand reduction.

RATE DESIGN

The primary purpose of any rate design is that it generates the necessary revenue for the utility in a manner that is fair and based on the cost of providing service.[82] James Bonbright and Charles Phillips, in perhaps the seminal work on utility rate structure, describe sound rates as those achieving:[83]

- Effectiveness in yielding the total revenue requirement
- Fairness in relation to the cost of serving different types of customers
- Practicality, including simplicity, understandability, ability to implement, and public acceptability
- Clarity in its interpretation
- Stability in revenues from year to year
- Continuity of rates, including the concept of gradualism
- Avoidance of undue discrimination among similarly situated customers
- Encouragement of efficient consumption practices

---

81   Ibid.

82   Scott J. Rubin, "What Does Water Really Cost? Rate Design Principles for an Era of Supply shortages, Infrastructure Upgrades and Enhanced Water Conservation," National Regulatory Research Institute, July 2010

83   James C. Bonbright, *Principles of Public Utility Rates Public Utilities Reports;* 2nd. ed. edition (March 1, 1988) (New York, 1961), 291; Charles F. Phillips, Jr., *The Regulation of Public Utilities: Theory and Practice,* Public Utilities Reports; 3 edition; July 1993 (Arlington, VA), 434–435

Throughout history various forms of rate structure have been developed and employed, not all of which satisfy Bonbright and Phillips's criteria. In municipal applications, water service has been provided under many differing cost recovery mechanisms, notably:

- Free
- Fixed Fee (no volumetric rate)
- Uniform rates (fixed base rate + static volumetric rate)
- Declining block rates (fixed base rate + decreasing volumetric rates)
- Pyramid rates (fixed base rate + increasing volumetric rates for small users and decreasing volumetric rates for larger users)
- Inclining block rates (fixed base rate + increasing volumetric rates)
- Water budget-based rates

FREE WATER

Clearly, free water is no longer an option, although it still exists in many jurisdictions. And in reality it has never been free to constituents because the costs of operating any municipal infrastructure are recovered through the revenue stream. The primary problem with free water is that it perpetuates the invisibility of water while simultaneously reinforcing the myth that water is ubiquitous and continuously available.

FIXED FEE

Prior to metering (and in some cases even with metering), water charges were assessed on a flat fee structure. All customers paid the same fees regardless of usage (e.g., $20/month). While this has the benefit of generating revenue for the utility and increasing the general public's awareness of water, there is no incentive for the customer or the utility to reduce usage.

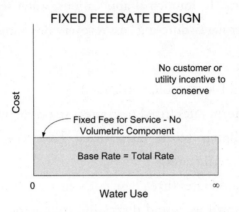

Figure 18: Fixed fee rate design

UNIFORM RATES

Uniform rates entail a small "fixed" fee with a nonvariable volumetric charge. This rate structure provides no incentive to conserve (that is, there are no additional charges for using more) and the small fixed component drives the utility to encourage consumption as a means to ensure revenue or even generate incremental revenue.

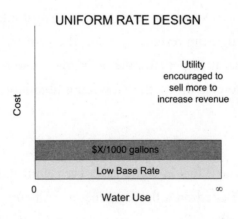

Figure 19: Uniform rate design

DECLINING BLOCK RATES

Declining block rates are employed to provide volumetric breaks to large customers under the premise that the marginal cost of providing large amounts of water to a small number of users is lower than providing the infrastructure, operations, and maintenance for a large number of lower-consumption users.

However, in many cases, large water users can actually define a system's necessary capacity (and thus the cost of the system's infrastructure.) As an example, if you consider that a large industrial customer may demand short periods of 2,000 gallons per minute, the water and infrastructure needs to be there for that service continuously. This is the equivalent of the peak hour flow from 3,400 single family dwellings.[84] Notwithstanding that the demand may be intermittent, the infrastructure needs to be there.

Regardless, the declining block model encourages the customer and the utility to increase consumption: the customer because his or her marginal cost declines with use, and the utility to ensure the revenue requirement is met.

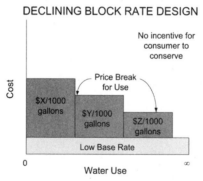

Figure 20: Declining block rate design

---

84   If the average annual demand from a single family dwelling is 250 gallons per day, the peak hour flow can be calculated as follows:

1. Average day flow = 250 gallons per unit per day
2. Maximum day flow = 495 gallons per unit per day (250 x 1.8 + 10% for potential line losses)
3. Peak hour flow = 0.58 GPM per unit (1.7 x max day flow)

PYRAMID RATES

Pyramid rates combine inclining block rate structure and declining block rate structures in an attempt to eliminate the detrimental aspects of each. In this case, customers are charged higher volumetric rates as consumption increases to a specified point, and then volumetric costs decrease.

Such a system does not promote conservation for the higher customer. Wang notes "it is likely not accurate, however, to consider pyramid block rates to be water-conservation-oriented rates, as they result in the highest consumers within the commercial class paying less per unit that those who use less."[85]

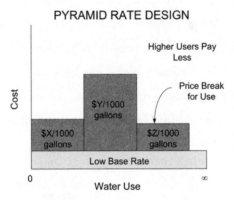

Figure 21: Pyramid rate design

INVERTED BLOCK RATES

With inclining block rates, more consumption means higher cost. While recognized as a water conservation rate structure, the inverted-block-rate (IBR) structure encourages the utility to sell more water at a higher price. In addition, IBRs are typically designed with low price

85   W. J. Smith, John Byrne, and Y.-D. Wang, *Water Conservation-Oriented Rates: Strategies to Extend Supply, Promote equity, and Meet Minimum Flow Levels,* (Denver, CO: American Water Works Association, 2005), 5

signals (that is, small tier increments) and broad tier ranges that lose effectiveness quickly in the consumption spectrum.

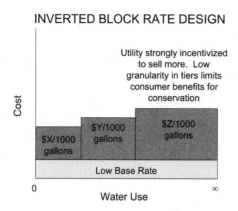

Figure 22: Inverted block rate design

CONSERVATION ORIENTED RATES

All standard rate designs suffer from significant shortfalls when it comes to encouraging conservation and ensuring the financial viability of our utilities. To be effective, rate designs need to be based on information—more to the point, they need to be as "personal" as possible to achieve real behavioral change. Two rate designs provide this—water budget based rates and rebate threshold rates.

WATER BUDGET-BASED RATES

Water budget-based rates define "normal" usage for a particular property based upon actual water use in relation to their customized water budget for the property. The customized water budget can include many factors including number of occupants, area of green space, evapotranspiration. Customers whose water use remains within their water budgets are billed at lower rates, and customers who exceed their budgets are billed at higher consumption rates.

The water budget is calculated for residential customers based upon surveys conducted of each customer's lot size and landscaped area, the number of residents in each home, and localized weather data. Water budgets change with seasons. For commercial applications, water budgets are established on rolling averages of use.

A key difficulty in establishing water budgets is the necessity for frequent and up-to-date property information with respect to population and landscaping, increasing manpower commitments to manage the program. However, when combined with a geotemporal data model and customer oriented messaging, water budget-based rates can be integrated into an effective conservation oriented rate structure.

Water budget rates can provide a mechanism for a community to focus water use on desired outcomes. For instance, if a community wants to encourage residents to create green space such as a tree-lined boulevard, it can increase the water budget for these properties. Conversely, a community focusing on xeriscaped, natural landscaping could reduce the water budget providing a financial incentive to change. The primary advantage of water budget rates is that they provide for a technical view of the actual water requirements necessary for specific properties under a directed management approach, and allow for both positive and negative incentives within the customer's control.

REBATE THRESHOLD RATE STRUCTURE

Today the altruistic desire to conserve is being overtaken by practical requirements. Water scarcity, dwindling historic supplies, environmental diversions, population growth, and regulatory requirements all conspire to send a clear message: use less water tomorrow and even less the day after that. However, the decisions that drive conservation are most often based on economic factors at the household level.

The Rebate Threshold Rate (RTR) structure is designed to reflect this reality and to maximize the behavioral change opportunities for customers.

This is achieved by the adoption of three basic elements in the RTR:

- Volumetric rebate and increasing volumetric charges in higher tiers to compensate;
- Increasing the number and granularity of tiers; and
- Achieving some measure of rate decoupling by increasing the fixed rate component.

Volumetric Rebate

The volumetric rebate allows for residential customers who achieve real, immediate reductions in water consumption to realize an immediate reduction in their volumetric charges. This process works by establishing a rebate threshold volume. Any time a customer achieves a consumption level below that of the rebate threshold, the customer is entitled to receive a reduction in volumetric charges (commodity charges). That reduction is typically 45 to 65 percent.

The rebate threshold can be established at 90 percent of the average residential consumption for a specific period, and remain static between changes in rates, but it could easily be tuned annually or as desired. As water elasticity occurs over longer timescales, it is important to allow people sufficient time to develop personal water management techniques and practices to maximize their benefit.[86]

---

86 A key supporting element of the RTR rate design is customer feedback with sufficient granularity for customers to make changes in near-real time to control their costs.

Volumetric Tiers

An increase in the number and granularity of tiers in the RTR structure allows customers to manage their usage, even if they are not below the rebate threshold, and still achieve meaningful cost reductions. Further, it ensures that there are greater financial disincentives as water use increases.

Today utilities typically employ a two- or three-tier rate structure in an inverted block design. The goal of the inverted tier rate is sound policy. The downside is that in limiting rate design to two or three tiers, those tiers are by necessity broad, limiting the ability for customers to achieve meaningful cost savings and reducing financial incentives to conserve. This in turn means that customers have fewer opportunities to manage themselves to a lower tier. The result is that fewer customers realize a true cost saving, and hence the incentive to conserve wanes. In addition, if the financial incentive for conservation ends far below a customer's consumption, there is no increased marginal cost to continue that behavior-modifying activity.

Increasing the number of tiers offers a number of "gates" through which the customer has the option of passing, or not, placing the control of the customer's volumetric costs squarely in the hands of the customer. In the case of a two- or three-tier system, the gates are passed too quickly and with little fanfare. The incentive to conserve through the traditional three-tier price points is lost after 10,000 gallons. With a six-tier design, customers have an incentive to think about different water price points through 25,000 gallons.

In a six-tier system, with the tiers established across effective thresholds, the customer has an opportunity, through active management, to maintain his or her consumption in a lower tier, and receive the benefit of the lower rate. Also with a six-tier system, finer modifications to rates can be achieved, saving customers money, and reinforcing the conservation message. Tiers can also be tuned to specific users, if desired, building on the "individuality" of the water budget-based rate structure.

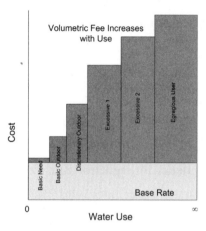

Figure 23: Rebate threshold rate (RTR) tier design

Revenue Decoupling

The monthly minimum charge (or base rate or basic charge) allows the utility to effect meaningful, measurable, and repeatable resource conservation without the implosion of utility revenue. Historically, support for conservation in the water utility business has been suspect: the utility knows that by encouraging its customers to use less, there is a real chance of revenue reduction, and even potentially a conflict with used and useful doctrines because infrastructure may be seen as "unnecessary" in the context of a reduced demand.

To achieve conservation goals, we must break the cycle of selling more water. By allowing for the recovery of fixed costs with a bias toward the monthly minimum, we can achieve both goals.

From a theoretical economic perspective, the provision of water service can be separated into infrastructure (fixed) and delivery (variable) costs. In practice, the dividing line is not as neat. Nor should it be, particularly as the goal of rates shifts from specific cost recovery to behavioral modification for the purposes of resource management.

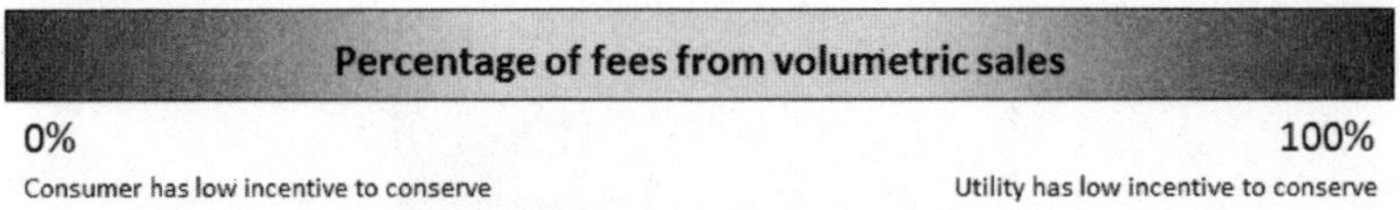

Percentage of fees from volumetric sales	
0%	100%
Consumer has low incentive to conserve	Utility has low incentive to conserve

Figure 24: Spectrum of volumetric component of rates

Attempting to definitively separate infrastructure and delivery costs can lead to some unintended consequences. As utility plants depreciate, there can be a tendency for regulators and city councils to reduce the "fixed" component of water rates (with the belief that full cost recovery has been achieved) and recover more utility costs through the "variable" or volumetric component. Under this condition, the utility loses its incentive to encourage conservation, because a reduction in use means an immediate reduction in revenue. Further, there is no incentive for the utility to invest in replacing aging infrastructure.

Clearly, if the bias is toward 100 percent cost recovery via monthly minimum charges and no increasing commodity rate, there is no incentive for the customer to conserve. Conversely, biasing rate structures to recover all costs via the commodity rate creates a strong economic disincentive for the utility to promote water conservation. By establishing a reasonable apportionment of costs to the monthly minimum and the commodity costs, both goals are achieved.

By generating approximately 65 percent of the revenue from monthly base fees, the financial viability of the utility can be assured. Further, the RTR design requires that the higher users contribute more as a function of consumption. In essence, the revenue shortfall for low users (those actively conserving) is made up from those users employing large amounts of water.

REBATE THRESHOLD RATE DESIGN

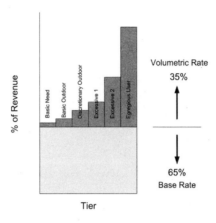

Figure 25: Rebate threshold rate (RTR) fixed vs. variable design

RATE TUNING: THE CARROT AND STICK

Conservation oriented rate structures allow utilities and regulators to tune rates to achieve specific objectives. For instance, the rebate threshold or budget can be incrementally lowered to drive consumption down; or the granularity of tiers can be altered to ensure that lifeline water supplies always remain within access for low-income households; or the utility can establish tier levels to derive more revenue from egregious users to fund conservation infrastructure projects; or the utility can fine tune the "gates" for individual users.

As an example, if a utility desires to fund conservation infrastructure (e.g., recycled water distribution systems, smart grid for water installation, etc.), it can increase the cost of higher tier water (tiers 4, 5, and 6) while maintaining lower tiers to ensure accessibility to water for all, while having those customers who use more pay a higher proportion of the costs of the provision of that service.

TUNING THE RTR STRUCTURE

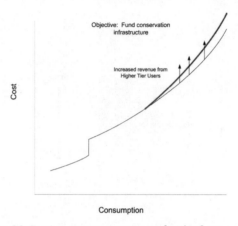

Figure 26: Tuning rate structures to fund infrastructure

Alternatively, a utility may wish to reduce overall demand due to dwindling resource availability. In this case, the utility can shift the rebate threshold to a lower volume, providing an economic incentive for all users to reduce demand. Users around the rebate threshold will adjust usage to maintain their economic incentive, and users above the threshold will adjust usage to remain in lower cost tiers.

TUNING THE RTR STRUCTURE

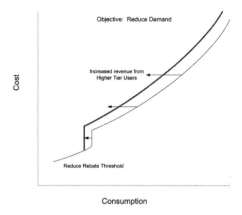

Figure 27: Tuning the RTR to encourage conservation

WILL OF THE PEOPLE

A key question is whether regulators or city councils have the will or the ability to execute on a strategy that both brings financial stability and conservation at the same time. As we know, the result of price increases will be public outcry, but does it have to be? As Scott Hempling of the National Regulatory Research Institute points out:

> Those years-long rate freezes lull the public into thinking rate stability is an entitlement. When, after ten years of below-cost rates, the commission realigns rates with cost, we know what happens: (1) Voters don't offer thanks for the prior windfall; they protest the new levels, loudly. (2) Politicians fan these flames, making rational policymaking difficult. (3) The compromise arrives, usually more pain deferral than pain sharing, thus skirting the underlying problem (the public's lack of acceptance that utility costs, like all costs, rise). What works in politics—mediating between positions—rarely works in

regulation, where the midpoint between two wrong answers is a third wrong answer.[87]

The problem lies in the fact that regulators (be they public utility commissions or city councils) do not regulate customers; they regulate utilities. And if conservation is a public interest goal, they must provide the incentives to customers *through utilities*. Part of that program is getting the prices and price signals right so as to eliminate waste, but also to serve the competing goals of customer protection, utility financial stability, and achieving water sustainability.

Hempling continues: "Rate design is the key to consumer protection. To moderate cost increases, we must moderate the demands that cause costs. Rate design offers the double anti-oxymoron: price increases are consumer protection, because (1) price increases change behavior, and (2) behavior change yields lower total costs."[88] For regulators, increasing rates is a reality. To mitigate the impact on customers, they must demand that rate increases are accompanied with the information necessary for the customer to actively manage their use.

RECENT INCREASES

The result of increasing treatment requirements, replacement of infrastructure and degrading source water quality, and recognition of water's real value is that prices are increasing.

In Queensland, Australia, water prices have doubled in four years. The cost of water increased from $0.97/kilolitre ($3.67/1000 gallons) in 2005 to $2.32/kilolitre ($8.79/1000 gallons) in 2009.

---

87 Scott Hempling, "Low Rates, High Rates, Wrong Rates, Right Rates," National Regulatory Research Institute, 2009, http://communities.nrri.org/web/guest/essays/-/wiki/Main/Essay+16/pop_up;jsessionid=99C346D8A9461DD22EDB6EFAE338F8F4?_36_print=true

88 Ibid.

According to recent submissions to the Queensland Competition Authority, costs are expected to rise between 7.5 percent and 23 percent in 2011–12.[89]

Water prices in the United States increased at twice the Consumer Price Index (CPI) rate from 1983 to 2010.[90] Prices are expected to increase at a rate greater than that of inflation for the foreseeable future.

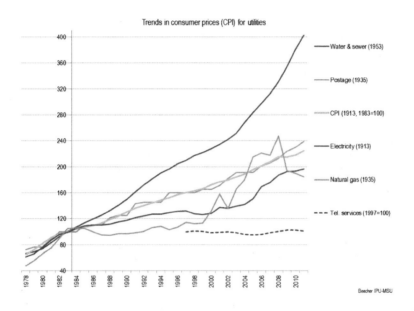

Figure 28: Increases in cost of utilities[91]

These price increases will result in increased interest and scrutiny from the "lay customer" for utilities, and the call for more information regarding consumption.

89   Courtney Trenwith, "Water usage dries up as prices rise," *Brisbane Times*, November 29, 2010, http://www.brisbanetimes.com.au/queensland/water-usage-dries-up-as-prices-rise-20101129-18dg8.html

90   Brett Walton, "The Price of Water 2012: 18 Percent Rise Since 2010, 7 Percent Over Last Year in 30 Major U.S. Cities," Circle of Blue, May 10, 2012

91   J. Beecher, "Trends in Consumer Prices (CPI) for Utilities through 2011", Institute of Public Utilities, Michigan State University, 2012.

WHAT MOST CITIES DON'T KNOW ABOUT THEIR HUMAN AND PHYSICAL INFRASTRUCTURE

As we know, there are known knowns; there are things we know we know. We also know there are known unknowns; that is to say we know there are some things we do not know. But there are also unknown unknowns—the ones we don't know we don't know.

—DONALD RUMSFELD, FEBRUARY 12, 2002

Water and sewer infrastructure suffer from invisibility. The infrastructure is in the ground, nonhazardous (compared to natural gas and power), and decidedly low-tech. Further, much of the infrastructure was installed not by city or utility crews but by developers and land owners. It is not surprising, then, that most utility departments, unless they have been judicious about requiring record drawings and as-built drawings, have very little knowledge of the location of their systems. Further, it is unlikely that utility employees have access to vital records such as piping material, size, installation dates, and so forth, which severely limits their ability to respond to both emergencies and routine calls.

An oft-cited quote in quality management is that you cannot manage what you do not measure. In the water business, you can't manage what you do not know is there.

Many utilities still rely on quarter-section maps, marked up with hand-written notes to identify infrastructure. Not only is this an inefficient data repository, it limits the availability of that data to one person. To be successful, the twenty-first-century water utility must become a "data gateway," eliminate data gatekeepers, and maximize the potential uses for data.

The implications of this invisibility are manifold. They go beyond infrastructure, through water, and directly to revenue and the customer experience.

NON-REVENUE WATER

Water lost is revenue lost. But it is not simply burst water mains; non-revenue water (NRW) can be a result of many factors:

- Leaks (both visible and invisible, those leaks that never reach the surface)
- Theft
- Metering inaccuracy (due to meter age or particulate precipitation)
- Meter programming inaccuracy (e.g., meters programmed to read in thousand gallons, but entered into the billing record as reading in gallons)
- Meter loss (i.e., meters missing from the billing inventory)
- Utility-consumed water (line flushing, sewer flushing, backwashes, etc.)
- Data transcription errors (manual data entry)
- Human errors

While the water manager may understand the volume of water delivered to, and sometimes from, the water distribution center, once the water is delivered to the distribution system, the water manager cannot tell where it went. Billing systems that operate on a monthly, bimonthly, or semiannual basis generate data relevancy problems that cannot be resolved: any problem that is identified at the billing data frequency provides little or no operational value. Further, identifying problems on a monthly basis risks leakage of revenue and water for a significant period of time before detection.

In response, water managers and utilities are turning to solutions that not only accurately read meters at a dramatically increased frequency, but also assist in finding and controlling leaks. Reducing non-revenue water (the difference between water pumped, treated, and supplied to the distribution system versus water that actually reaches customers) is of critical importance for the financial and hydrological health of the utility, and resource sustainability.

Increasingly, the answer lies in improving the temporal and spatial quality of the data. For example, by combining highly granular consumption data with addresses from the customer information system and georeferenced spatial data, a rapid, visual identification of water theft can be built, allowing operations staff to take action.

VANISHING WORKFORCE

Alongside water, utilities are losing corporate knowledge at an alarming rate. Overall, the utility workforce is aging. Nationwide, 22.5 percent of utility workers are at or above the age of fifty-five.[92] For the water sector, however, the problem is more acute with 70

---

92   US Census Bureau, Local Employment Dynamics, http://lehd.did.census.gov/led/datatools/qwiapp.html, data for North American Industry Classification Systems (NAICS) 221 utilities

percent of states (thirty-two of forty-six[93]) indicating that the percentage of water employees at or above the age of fifty-five is greater than the national utility average.

In addition, the overall age distribution of utility workers is significantly skewed as compared to the workforce in general, with a larger proportion of the industry moving into retirement age.

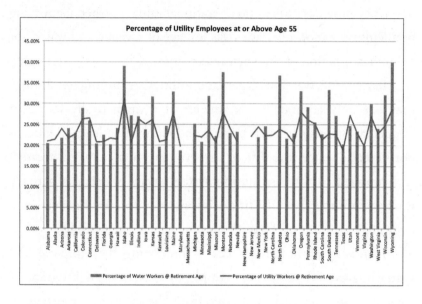

Figure 29: Utility worker age by state

93   US Census Bureau, Local Employment Dynamics, http://lehd.did.census.gov/led/datatools/qwiapp.html, data for NAICS 2213 utilities, water, sewage, and other systems. Massachusetts, New Hampshire, New Jersey, and North Carolina do not report NAICS subcategory 2213.

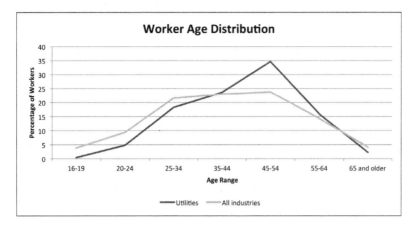

Figure 30: Utility worker age distribution

This represents a significant challenge for utilities: how to retain and store a generation's worth of knowledge about systems, so that it remains available and easily accessible. The problem is exacerbated by the fact that the operation of water systems is a knowledge-based industry that necessitates significant training and experience to be effective. As noted in a recent C. D. Howe report on ensuring public safety:

> The first step is to recognize that providing safe drinking water is a knowledge-based activity. This activity cannot be downloaded to the same level of municipal priority as garbage collection and snow removal. Those assigned to provide drinking water have to be afforded the training, intellectual support, and compensation that is commensurate with taking responsibility through their actions or inactions for the health of an entire community.[94]

---

94   Steve E. Hrudey, "Safe Drinking Water Policy for Canada: Turning Hindsight into Foresight," C.D. Howe Institute, *The Water Series*, no. 323 (February 2011), 18.

Ensuring that operations staff members "know their systems" is a critical facet of the provision of safe drinking water.

Maintaining the knowledge necessary to ensure public health will be an ongoing challenge for utilities.

A CUSTOMER INFORMATION SYSTEM IS NOT A WATER MANAGEMENT TOOL

In God we trust; all others must bring data.

—W. EDWARDS DEMING

Effective water scarcity management is a utility-wide, knowledge-based activity. Unfortunately, most data systems in utilities operate in departmental silos, designed, built, and operated for their specific objective, with very little cross-functional use of the data. Even in advanced utilities, customer information systems (CIS), supervisory control and data acquisition systems (SCADA), billing platforms, asset management systems, geospatial information systems (GIS) are often separate, managed by different departments, employed for specific reasons, and incapable of easily sharing data.

These disparate data systems are a cause of significant frustration for utility managers, but also contribute to significant losses in efficiency. No doubt all readers can relate to the tale of exporting data from one system, coaxing it through an intermediary platform (such as Excel or Access) and importing into another system. The time spent getting the data, cleaning the data, processing the data and reformatting it for use in another system is simply a resource sink.

Further, the granularity of the data employed in these systems can differ by orders of magnitude. SCADA data is available at millisecond intervals; customer consumption data is available at monthly intervals. This timescale disparity complicates the ability to effectively use the data.

And these disparate systems are employed by disparate groups within the utility. Rarely will the billing manager access SCADA data. Similarly, to the operations manager, monthly consumption data provides no basis from which to make decisions. Although, provided with the correct timescale relevance, each could arguably benefit substantially from each other's data.

While other industries have increased the speed, volume, and availability of data in order to maximize efficiencies, many water and wastewater utilities continue to manage water resources and revenue based on singular, isolated data points. The result is that the impact of management changes (in operations or resource availability) cannot be detected for months, slowing any progress to efficiency. Worse, changes in conditions that can affect the financial viability of a utility cannot be predicted or discerned until it is too late.

DRIVERS FOR DATA INNOVATION

The benefits of increased data availability and speed are manifold and varied. However, the primary drivers for this innovation can be reduced to several fundamental challenges existing in water management today:

- Increased awareness of water scarcity, requiring real-time knowledge of the source, distribution and use of all water.
- Increased pressure on municipal finances, requiring maximum efficiencies to be developed.

- Increased awareness of price signals and customer impact requiring the provision of information in order that customers understand the reasoning and impact of price increases.
- The understanding that data can dramatically improve systems maintenance and emergency response, and ensure public health is maintained.

The future for efficient utilities lies in collecting and analyzing this data, and converting it in real time to information. To be successful, we will need to increase our data awareness by:

- Increasing the amount of data available to the utility and its customers
- Increasing the capability of our systems to manage and analyze this data
- Increasing our ability to present this information in meaningful ways to shape behavior patterns of customers and the industry.
- Interconnecting data and systems is evolving into the smart grid for water. By increasing the availability of data and information, the smart grid for water can tell us not only *how much* water is used, but *where* and *when*.

SMART GRID FOR WATER

The smart grid for water represents the new reality for water utilities. Originally founded on the same basis as the electrical smart grid, the basis for investment in the smart grid for water is different. For electrical utilities, smart grid and user-driven conservation is a means to defer the construction of generation capacity. By providing the data and impetus necessary (from a rate and consumption perspective), electrical utilities can alter the demand response so that more power can be shifted to "off peak" times, thus reducing the peak load demands. This has some relevance in the water utility business as we shall see later, but for most, the smart grid for water is about revenue assurance, responsible water resources management, and water asset management.

The smart grid for water includes the following components:

- Advanced metering infrastructure (AMI)—electronic meter reads and communication backbone
- Customer information systems—billing and account management
- Supervisory control and data acquisition systems—equipment control
- Computerized maintenance management systems—work order generation and control
- Geographic information systems—spatial referencing and data management
- Analytics engines

One of the first realizations about the smart grid for water is that it is not about meter end points and communications. While these are important considerations, the smart grid for water is a data analysis platform, not a data collection platform. Unfortunately, many automated meter reading (AMR) systems that purport to provide a smart-grid-for-water solution, in fact only collect the data and leave the analysis out of the equation.

It's for this reason—data integration and analysis—that pilots are seldom indicative of the activation energy required to complete a smart-grid installation. Pilots are generally "hardware-centric," ignoring the much more complex software integration requirements. Many organizations chose the AMI hardware provider first and then attempted to integrate the hardware into the software in the enterprise. This is backward. The real benefit to a smart grid solution is data and how that data is integrated into the business platform. Having a solution driven by the meter end point is a serious flaw. It is much better to build process and practice supported by hardware than the reverse.

Smart grid for water installations need to be designed from the perspective of the use we expect to develop from the data, and then worked upstream to define the data source necessary for those decisions. The smart grid for water should really be hardware and software agnostic provided the hardware and software can meet the needs of the business platform and practices, and the platform synthesizes the data into information.

The risk of obsolescence is a real factor. As many utilities have found in the past, selecting legacy billing or other software platforms can lead to significant costs in the future, or seriously hamper future data plans. As a result it is important to consider the longevity, or "evergreen" nature of any platform.

INCREASING DATA DENSITY

In the past, water utilities employed a data-poor model upon which to make decisions. One read per month, or worse, one read per quarter. Automated meter reading (AMR), while notionally increasing the data density available to the utility, in many cases only supplanted the meter reader. The result: one read per month. No increase in available data, but timeliness and accuracy were arguably increased.

Moving to a true "data-rich" model requires advanced metering infrastructure (AMI), with its attendant communication backbone and increased frequency of reads. When combined with geospatial referencing, the data model achieves the data-rich capacity of the geotemporal data model.

The transition from a data-poor condition to a data-rich environment not only provides better information upon which to base decisions, but increases the speed at which those decisions can be made. For example, by deploying a system that can provide real- or near-real-time, water-pumped versus water-delivered reports, a leak can be recognized immediately. Water theft can be identified rapidly. Missing meter data can be noticed instantly.

But such data-rich systems are a double-edged sword. The vast quantities of data accumulated can easily overwhelm IT systems and the human capacity to assimilate. Typically, it is the pitfalls of data management and analysis that hinder the successful integration of utility data systems into the smart grid for water and prevent the transformation of these databases into a coherent *information system*.

Utilities often struggle with this AMI data deluge. Management systems designed for one meter read per month must now deal with 720 reads per month. A "simple" meter read is now represented by a data stream that is three orders of magnitude above what many systems can handle. And it does not stop there. When additional data such as leak/tamper flags, time/date stamps, peak/average flow tags, and so forth, are factored in, the data volume increases exponentially. As with many data collections systems, in the smart grid for water, the bottleneck has shifted away from the data collection to data analysis.[95]

In this data flow, the benefits of the smart grid for water data can easily vanish, causing the users of the system to stem the tide. In

---

95   Richard G. Baraniuk, "More Is Less: Signal Processing and the Data Deluge," *Science* *331* (February 11, 2011): 717

these cases, data collection is "dumbed down" to the lowest common denominator of the systems. This could be a limitation on the customer information or billing systems platform (e.g., systems can only accept one read per month), or a limitation on storage capacity (physical or virtual), or a limitation on the ability to comb through the data to create information (analytical capacity and capability). In these cases, the increased data availability becomes a curse, and the response can be to rapidly reduce data collection from 720 reads per month back down to one. The utility has just inadvertently bought itself a very expensive meter reader.

INCREASING DATA ANALYSIS

The smart grid for water has been erroneously defined as the AMR/AMI system, that is, focused around the "hardware-centric" elements of the system, the end point and the communication infrastructure. But it is and will increasingly prove to be much more. The power of the smart grid for water is in the convergence of hardware, data, and software to collect, search for, and identify trends in data streams, and maximize the efficiency of water delivery, customer service, and utility operations.

Some of this analysis is already embedded within the AMI systems or meters themselves. For example, leak detection and tamper flags can be processed signals that are derived from raw data in the end point or meter register.

But the real benefits of the smart grid for water lie in aggregative, integrative, and derivative information that can be gleaned from combining data, particularly from those three critical databases that all utilities have, or should have: Customer information systems databases answering the question of who, coupled with financial information; advanced metering infrastructure/SCADA databases

87

answering the questions of how much, and when; and geospatial databases answering the question of where our assets are and where use occurs.

In short, reviewing data from numerous sources and developing tangible, actionable information.

Under this approach, a meter read is not just a meter read. It forms a key part of the billing record; it forms a fundamental part of the leak loss (pumped versus billed, or non-revenue water) analysis; it establishes peak and average demand parameters; it is a key measure of the performance of water conservation activities; it forms the basis for feedback to the customer directly on their impact on resources; and it is the foundation for key reporting elements associated with regulatory requirements such as compliance with California's 20x2020 Water Conservation Plan.

The key is that while this information is aggregated and derivative, the data source itself is singular. There is no need to store this data point across multiple platforms. Gather the data once; use it many times. A recent International Data Corporation report notes that "nearly 75% of our digital world is a copy—only 25% is unique," and "the greatest challenges are related not to how to store the information we want to keep, but rather [in] extracting all of the value out of the content that we save."[96] A properly deployed smart grid for water provides the analytical tools to solve this dilemma.

INFORMATION PRESENTMENT

Having masses of data and the algorithms to reduce it to information are only as good as the tools provided to people to action the information. Information presentment represents the kernel of actionable

---

96   John Gantz and David Reinsel, "The Digital Universe Decade—Are You Ready?" IDC document #925 (May 2010): 9, http://idcdocserv.com/925

information from the data: something that can be done; a change made. It can take the form of instantaneous water-pumped versus water-billed reports reconciling flows on a daily basis; or maps of unauthorized water usage, or egregious water users, or super-conservers; or demonstrating to customers their water consumption versus their neighbor's or their community; or allowing them to track daily their gallons per capita per day (GPCD), on a year-to-date basis.

But these information presentment tools are often ignored. As the volume of data increases, "the extra effort of making our data understandable, something that should be routine, is consuming considerable resources."[97] And consumers of data, be they utility customers or utility managers, are increasingly conditioned to operate in a data-rich world. Failure to recognize the importance of data presentment is a fatal flaw in a smart grid for water deployment.

97   Peter Fox and James Hendle, "Changing the Equation on Scientific Data Visualization," *Science 331* (February 11, 2011): 705

CHANGING UTILITY DATA SYSTEMS TO INFORMATION SYSTEMS

By 2045, [the] non-biological portion of our civilization's intelligence will expand a billion-fold. Our biological intelligence is very impressive, but it's fixed. It's not going to expand. So, ultimately, we'll be dominated by non-biological intelligence.[98]

—MCKINSEY QUARTERLY (JANUARY 2011),

"IT GROWTH AND GLOBAL CHANGE:

A CONVERSATION WITH RAY KURZWEIL"

Water utilities have not kept pace with the increased volume of data nor the demand for information. The nineteenth-century data solution still pervades the majority of water utilities. One-dimensional data are insufficient for managing in today's environment. Single consumption points misaligned with billing and revenue time frames confuse rather than inform, and attempting to use this data to infer complex relationships between demand and supply, seasonality, and variability of revenue as a function of conservation can be at best frustrating and at worst futile.

---

98   *McKinsey Quarterly* (January 2011), "IT Growth and Global Change: A Conservation with Ray Kurzweil"

Change is coming, however. Not surprisingly, these changes are being driven by external factors, most notably water scarcity, and the financial condition of utilities. But despite the negative impetus, the potential benefits are broad. Providing the twenty-first century water resources manager with real-time actionable information through increasing the availability, frequency, and velocity of data needed to make decisions allows for better, faster, and more robust responses to external forces and internal conditions. Combining data platforms and deploying data mining algorithms allows for significant decision support systems to be established, fundamentally altering utility management practices.

To achieve all the benefits of such a knowledge-based management model, a new data model must be employed. Disparate data systems operating in isolation must be eliminated and their data integrated. The granularity of data must be increased. The time basis of the data must be known. And the data's spatial relationship must be tracked.

Geospatial information systems (GIS), customer information systems (CIS), computerized maintenance management systems (CMMS), supervisory control and data acquistion systems (SCADA), financial systems, and other data sources and users must be brought together to allow for the generation of knowledge. This new data model is the source for decision support tools that allow for the maximum "knowledge efficiency" in an organization.

To get there requires a quantum change in the way we view data, the way we collect data, and the way we interact with data. Data can no longer be considered a static element in the utility whose charge it is to manage water and financial scarcity, nor can it be limited to use in arrears. The full force of data must be brought forward in time to be useful. It must be placed within the context of space and time. This is the geotemporal data model for utilities. Not only is it important to understand *how much* water is used, but *where* and *when*.

Dynamic and flexible links to all data sources must be established, including AMI, SCADA, GIS, CMMS, CIS, and other data platforms:

> [Utilities] no longer look at AMI as just faster meter reading; they want a network that provides the ability to tie in all these components. From water coming into the utility, to customer delivery, distribution lines, measurements at customer sites, and then allowing access to that data so the consumer becomes more intelligent on their consumption.[99]

The geotemporal data model provides the fastest, most up to date, and most accurate data available. Such data-rich environments also allow for the development and deployment of advanced data analysis tools, and for the rapid validation of data models to develop decision support systems. Genetic algorithms, neural networks, auto-regression analyses, statistical process controls, and expert systems can all be developed with the density of data gathered in the geotemporal data model.

Supported by well-defined business processes guiding the optimization of systems, the geotemporal data model forms the basis of the utility of the future. The geotemporal data model is a fundamental aspect of the smart grid for water and indeed the benefits of either cannot be realized without the supporting elements of the other.

---

99   Ed Ritchie, "A New Paradigm: Demand Forecasting and the Art of Resource Management," *Water Efficiency* (November–December 2010), http://www.waterefficiency.net/november-december-2010/data-integration-resource-management-software.aspx

HIDDEN REVENUE IN
THE SMART GRID

The smart grid for water is about analysis: analyzing data for managing scarce water resources, analyzing water accounting, both financial and resource.

A key benefit of increasing data availability, particularly through AMI installations, is the impact on utility revenue. While stopping water leaks is an important aspect of utility operations, stopping revenue leaks is vital to the survival of the enterprise. Thomas Chesnutt of A & N Technical Services notes that "forecasting and revenue were on the minds" of many attendees of the American Water Works Association (AWWA) 2010 Annual Conference and Exposition. "The big buzz around town is what people are doing about revenue, because there have been some ugly surprises."[100]

WATER REVENUE MANAGEMENT:
BILLING EVERY DROP, STOPPING THE REVENUE LEAKS

Utilities need to monetize every drop of water; water lost is simply revenue lost. And since water is a consumable commodity, once the opportunity to measure and capture the volume data has passed, it is impossible to retreat and recover this lost revenue. For the utility,

---

100 Ed Ritchie, "A New Paradigm: Demand Forecasting and the Art of Resource Management," *Water Efficiency* (November–December 2010), http://www.waterefficiency.net/november-december-2010/data-integration-resource-management-software.aspx

this means an updated mechanism for understanding distribution in real time is required, a mechanism that tracks the molecules from source to use (and then back and reuse). At a minimum, accurate and timely pumped-to-billed reports are required, but equally important is the requirement to understand whether all customers are actually being billed.

SCADA is capable of providing very accurate and timely reports on the amount of water operations are putting into the system and taking out. The same cannot be said of the meter/billing system. The single-dimension aspect of billing data eliminates the ability of managers to notice, correct, and recover from any water missed.

Similarly, billing systems and connections grow throughout the life of a utility. As these systems grow, it is not uncommon for the electronic records used for billing and the physical installations on the ground to become decoupled. The result is that some connections never make it into the billing system, which obviously means no revenue, increased water loss, and frustration for the utility manager.

The smart grid for water alleviates these issues by ensuring:

- All the meters are read all the time
- Eliminating the expense associated with missed and/or read errors
- Eliminating estimated reads

One of the key benefits of a properly deployed smart grid for water system, however, is that it forms the basis of an electronically validatable meter inventory. By exploiting GIS integration, the installation of each AMI end point is documented with GPS coordinates and georeferenced photography. This dataset can be cross referenced with city parcel records and the customer information system (CIS). The result is, for the first time, a full census of all meters and accounts.

Not only does this provide the opportunity to identify any meters that are not in CIS, but going forward with installation data from the field, combined with parcel and CIS data, provides an "electronically verified" installation, permanently eliminating the potential for lost meters.

OTHER REVENUE ENHANCEMENTS OF
THE SMART GRID FOR WATER

The financial benefits of a smart grid for water installation are numerous:

- Decreased labor expenses
- Real-time pumped-to-billed analysis
- High consumption/leak detection
- Water loss identification
- Water theft identification
- Allows errors to be determined precisely for back billing purposes

In areas where meter replacements are made at the same time as AMI installations, a significant source of additional revenue can come from the more accurate reads associated with the new meters.

ELIMINATING INEFFICIENCIES

The smart grid for water also provides the backbone for significant utility optimization programs, including:

- Development of a power optimization plan for water distribution
- Maximizing efficiency by understanding the real time peak and average demand

- Regulatory bylaw enforcement (e.g., water restrictions)
- Allowing for the integration of ancillary devices for increased efficiency:
  - Pressure sensors for outage monitoring
  - Pressure sensors for power efficiency (e.g., reducing pressure to reduce power)
  - Acoustic leak detectors for leak control
  - Smart irrigation controllers to better utilize outdoor water use
  - Real-time, hydraulic-flow modeling validation

IMPROVING THE CUSTOMER EXPERIENCE

Deploying the smart grid for water also provides for a significantly improved customer experience. Specifically, a smart grid for water installation can:

- Allow customers to understand their actual consumption in real time
- Provide high consumption alerts
- Provide leak alerts
- Allow for the benefits of price signals to be effective (Water prices are increasing along with customers' understanding of their consumption and the need to provide data to allow for active management.)
- Allow customers to actively control consumption
- Allow customers to set personalized alerts for consumption or billing

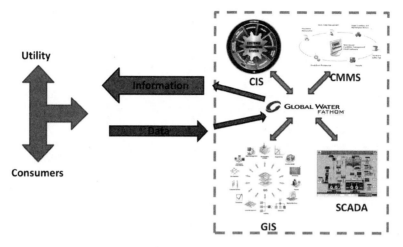

Figure 31: The smart grid for water

REAL-TIME DATA REQUIREMENTS TO CHANGE BEHAVIOR

For customers to embrace water stewardship, they need to be provided with a clear understanding of their consumption. They need relevant, highly granular, near-real-time data. Customers need the opportunity to review daily consumption, and make an economic decision based on that information. But equally important is data presentment and informal social influences. To be successful, a conservation program must get the data out to customers and make the change financially beneficial to the customer. But even more than that, people must be given the "geotemporal" context of their consumption:

- How much water do I use?
- How do I fare compared to my street, my neighborhood, my city?
- How much water should I use?
- Based on weather data and evapotranspiration calculations, how much should I have used outside?

Robert Cialdini, a psychologist at Arizona State University, recently noted: "People don't recognize how powerful the pull of the crowd is on them...We can move people to environmentally friendly behavior by simply telling them what those around them are doing."[101]

A recent study completed by California State University indicated that through the provision of instantaneous feedback on water consumption, average water consumption reductions in the order of 14 percent can be achieved.[102] And as noted by Paul Ormerod, engaging the customer with comparisons to neighbors is a very effective means of behavioral change: "Throughout history, a crucial feature of human behavior has been our propensity to copy or imitate the behaviors, choices and opinions of others."[103]

Combining price structure and appropriate signals, financial incentives, information, and subtle societal pressures, customer behavior can be altered to the point where he or she asks the questions: "Do I need to use that gallon of water?" and "Am I prepared to jeopardize my financial incentive to use the next gallon of water?"

To be effective, water conservation needs to engage the customer at a fundamentally more granular level than is achieved today and at timescales that make water utilities blush. While the industry has adopted SCADA systems to achieve highly optimized system-level control and automation, inexplicably, the same philosophy has not been applied to a massive area of the utility systems: customers.

Whereas a utility typically can tell from SCADA how much water is entering the distribution system instantaneously, billing platforms work at a much reduced timescale, eliminating any potential opportunity for real-time customer engagement.

---

101 Stephanie Simon, "The Secret to Turning Consumers Green," *Wall Street Journal*, October 17, 2010

102 Wesley Schultz, Warren DeCianni, and Alexis Roldan, "Water Conservation Pilot," research report, California State University, San Marcos

103 Paul Ormerod, "Social Networks Can Spread the Olympic effect," *Nature 489* (September 20, 2012): 337

The reason for this is easy to see. A utility, faced with a fixed real and in most cases highly regulated revenue stream, is incentivized to minimize costs to liberate funds for other projects. Therefore controlling equipment, optimizing power and chemical use, and eliminating unnecessary labor are critical elements. The result is investment in SCADA.

Conversely, for customers, the cost of water has rarely registered in their monthly budget. The result: no investment in customer-based consumption tools.

That's changing. American Water Intelligence reports that water and wastewater rates increased an average of 8.1 percent between July 2010 and July 2011[104] and are increasing at more than twice CPI rates. And price increases certainly make people more aware of their usage. In 2009 Boenning and Scattergood noted, "as with the sudden interest in fuel-efficient cars in the U.S., consumers generally become much more interested in conservation and efficiency when the price of the commodity in question—be it oil or water—becomes high enough to provide that motivation."[105] In other words, when water is cheap, no one notices. As prices increase, people look to control their consumption, which is good for water resources, but in order to affect that control, they will demand to know how much they are using. This is information that cannot often be provided with sufficient granularity to be a successful driver.

However, given access to highly granular, time-relevant data, customers can make dramatic changes in consumption.

---

104 American Water Intelligence, "Cities Hike Water Charges as Financing Options Evaporate," *AWI 2, no. 9* (September 2011)

105 Bill Malarkey and James Adducci, "Water Industry Review: M&A Market Outlook," (West Conshohocken, PA: Boenning & Scattergood, 2009)

THE DATA GATEWAY

Customers have become increasingly aware of, and demanding of, data. Compounding the data revolution are the changes in the way that people are communicating and interacting with each other, corporations, and systems. The world is becoming increasingly connected and wirelessly interconnected. "People everywhere are consuming more and more wireless bandwidth to manage a wider variety of tasks."[106] Real-time data is becoming a customer-imperative: customer demands for data pushed to a handheld device are on the rise, for example, satellite imagery and maps, GPS coordinates, location-specific and general news, sports results, and so forth.[107]

Indeed the volume of mobile data being generated and consumed is increasing exponentially and is expected to continue to do so:

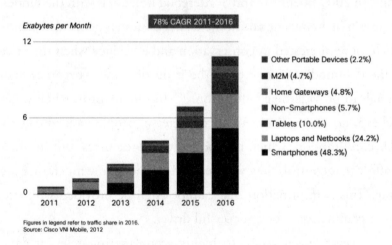

Figure 32: Worldwide mobile data use[108]

---

106 Evan I. Schwartz, "The Mobile Device Is Becoming Humankind's Primary Tool," *Technology Review* (November 29, 2010)

107 Madhavendra Richharia and Leslie David Westbrook, *Satellite Systems for Personal Applications: Concepts and Technology* (New York: John Wiley & Sons), 15.

108 Courtesy of Cisco Systems, Inc. Unauthorized use not permitted. "Cisco Visual Networking Index: Global Mobile Data Traffic Forecast Update, 2011–2016", http://www.cisco.com/en/US/solutions/collateral/ns341/ns525/ns537/ns705/ns827/white_paper_c11-520862.pdf. Accessed 31 January 2013.

This represents a fundamental shift in the way that people are interacting with data. Ignoring this shift to portable, instantly available data represents a substantial business risk, and a risk to maintaining high levels of customer service. Jesse Berst recently opined that "consumers want highly personalized information and they want it at any time on any device—Web, TV, print, smart phone."[109] It is for this reason that the information presentment options are converging to mobile devices such as Global Water FATHOM's iPhone application.[110] This application is one of the first commercially available water conservation/consumption data delivery tools for customers, and allows utility customers to access near-real-time consumption data and billing information.

There is tremendous power in this data. While traditionally slow to adopt and exploit the demands and availability of data, water utilities are awakening to the possibilities. This realization of the value of the data is a key shift in water management practice and customer consumption: "Understanding the role of information and the household consumer is integral for transforming a 'Water Supply City' where the focus is on infrastructure alone to a 'Water Sensitive City' where infrastructure, users and the environment are integrated."[111]

Jacques Bughin of McKinsey & Company notes, "Data rates are doubling every 18 months…the[se] new approaches help companies

---

109 Jesse Berst, "The Six Things Utilities Still Don't Get about Consumers (but Better Learn Fast!)," *Smart Grid News*, Apr 5, 2011

110 http://itunes.apple.com/us/app/fathom/id463933218?mt=8

111 Damien P. Giurco, Stuart B. White, and Rodney A. Stewart, "Smart Metering and Water End-Use Data: Conservation Benefits and Privacy Risks," *Water 2010*, 2:461–467, doi:10.3390/w2030461 ISSN 2073-4441 www.mdpi.com/journal/water

make decisions in real time. This trend has the potential to drive a radical transformation in research, innovation, and marketing."[112]

The twenty-first century utility manager must maximize the use of this data and transform it into information in order to ensure the efficient use of all resources—water, people, systems, and finances.

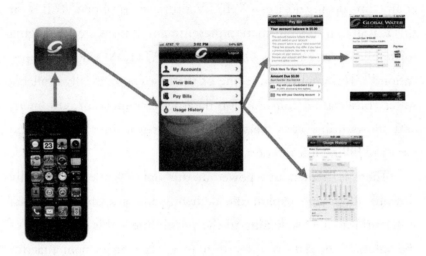

Figure 33: FATHOM mobile app

Ed Ritchie described a new paradigm in water information and communication technology.[113] Within this new environment he predicts that utilities are looking for systems that allow customers to access data "so the consumer becomes more intelligent on their consumption." The technologies are available and when integrated into a smart grid for water plan, provide a real and tangible connection for customers to their water.

---

112 Jacques Bughin, Michael Chui, James Manyika, "Clouds, Big Data, and Smart Assets: Ten Tech-Enabled Business Trends to Watch," McKinsey & Company (2010).

113 Ed Ritchie, "A New Paradigm: Demand Forecasting and the Art of Resource Management," *Water Efficiency*, November–December 2010, Volume 5, Number 6

ARE YOU LEAKING DATA?

Today's utilities are caught between the converging pressures of fiscal austerity, water scarcity, aging infrastructure, and increasing expectations from customers and regulators. Each of these presents significant challenges for utilities and, in fact, are conspiring conditions that threaten the vitality of the water industry.

As enterprise operations, utilities are typically required to generate sufficient cash flow and reserves to fund operations and their capital programs. In today's environment, however, there is increasing pressure to provide funding to general fund activities, and to reduce costs so as to improve the financial health of municipalities. That is particularly difficult, given the nature of the water business. Customers have been lulled into the belief that cheap, plentiful, and clean water must be available at all times. The result is that customers have been insulated from everything, from the cost of the molecule to the cost of maintaining the invisible infrastructure of the utility. The American Society of Civil Engineers estimates that by 2020 the capital infrastructure funding gap for water and wastewater will be $84 billion ($144 billion by 2040).[114] Closing this gap will take concerted effort.

---

114 American Society of Civil Engineers, "Failure to Act: The Economic Impact of Current Investment Trends in Water and Wastewater Treatment Infrastructure," 2011, http://www.asce.org/uploadedFiles/Infrastructure/Failure_to_Act/ASCE%20WATER%20REPORT%20FINAL.pdf

The current financial condition of many utilities—shrinking population bases, dwindling budgets, infrastructure replacement—mean that the utility must seek efficiencies well beyond the norm in order to achieve both continuity of revenue, and to fund future needs. While rates are increasing with time,[115] and this can be a means to fund budget and infrastructure shortfalls, these increases come with significant political costs. There is very little appetite for massive rate hikes, however imperative and appropriate they may be. The utility therefore, must seek to fill this financial shortfall through efficiencies.

One efficiency yet to be fully exploited by utilities is monetizing the entire utility water cycle. Ensuring that all water produced, treated, pumped, and distributed is actually delivered and actually billed is vitally important, environmentally and fiscally. That may sound like a trivial expectation, and on first blush it is. The reality is much murkier. The difference between the volume of water sent into the distribution system and the amount of water actually billed (non-revenue water, NRW), is a significant drain on utility revenue. The USGS has reported that US water systems experience 240,000 water-main breaks annually,[116] resulting in the loss of 1.7 trillion gallons of water every year. The USEPA reports that large utility breaks in the Midwest increased from 250 per year to 2,200 per year during a nineteen-year period, illustrating the fact that our infrastructure is nearing the end of its service life. In 2003 the City of Baltimore, Maryland, reported 1,190 water-main breaks, an average of more than three per day.

An equally important component of NRW is not leaking drops, but leaking data. The fundamental business tools used to ensure the

---

115 American Water Intelligence reports an average increase of 8.1% between July 2010 and July 2011.

116 USEPA, "Addressing the Challenge through Innovation," Office of Research and Development National Risk Management Research Laboratory, EPA/600/F-07/015, September 2007

utility's financial health are decoupled from the physical infrastructure. The result is that many utilities are not only leaking physical water, but are also leaking the data associated with the production, treatment, distribution, and selling of that water. And that means utilities are leaking dollars.

LEAKING DROPS AND DATA

Finding and stopping physical leaks is a critical aspect of utility management. The typical response to NRW is to deploy teams of distribution system technicians to track down, find, and repair the leaks. This may require the use of specialized equipment, or simply the specialized senses of the technicians. While we can easily comprehend the physical nature of water leaking from a pipeline, physical water leaks are only one of a myriad of potential causes of NRW: What if the problem is not really attributable to leakage? Or what if at least some of the water is actually staying in the pipes, but is not being measured? That is, the water is *leaking not from our pipes but from our data*. The actions of the teams of technicians will be futile.

To correct this problem, utilities must move to a smart grid for water installation, that is, advanced metering infrastructure with substantially increased data granularity and direct integration with customer information systems under a geotemporal data model.[117] Such a system will plug leaking data from time disparity, billing system errors, and meter degradation.

TIME DISPARITY OF SYSTEMS

One of the major problems in utilities is the time disparity within information systems. Tracking water requires detailed mechanisms capable of following a molecule through the distribution system

---

117 A geotemporal data model combines georeferenced physical installation information with highly granular time of use data.

(source to use to reuse). Unfortunately, a major component of the data necessary to understand the delivery water is not available, at least not at the timescales necessary for active management.

A utility can tell you to the millisecond when a booster pump turned on. However, in many cases it cannot tell you until next month, or the month after, or six months later, or in some cases never, where that water went. That's not acceptable. In a world where every drop counts, and every drop lost is dollars lost, instantaneous understanding of the entire water system is required.

We may, with good reason, decry the loss of the 1.7 trillion gallons of water each year noted by USGS for environmental and sustainability reasons. But the real tragedy in these statistics is that utility customers paid for acquiring, treating, and pumping that water. In some cases, where that treatment involves arsenic removal, or uranium removal, or ultrafiltration, or reverse osmosis treatment, the cost of creating that water to the utility is staggering. The EPA has estimated the cost of NRW to be in the order of $2.6 billion per year.[118] Finding this revenue can start to close the $84 billion infrastructure gap.

A smart grid for water installation allows utilities to track in near-real time where exactly each molecule of water is. With a system designed to provide hourly consumption information, utility managers can, for the first time, perform daily reconciliations of water accounting and identify the leaking drops early, and often, protecting revenue.

BILLING SYSTEM INADEQUACIES

Correcting this problem is a function of the utility's physical and information infrastructure. While SCADA is capable of providing

---

118 http://www.epa.gov/awi/distributionsys.html

very accurate and timely reports on the amount of water put into the distribution system, the same cannot be said of the meter/billing system.

In fact, meter reading/billing systems are, in many cases, the most antiquated IT platform in a utility. The reason is obvious: they are large, cumbersome legacy systems, too critical to be cracked open by the faint of heart. And most utilities lack the specialized skills to update and maintain these systems. The result is most utilities continue to get by on one manual read per month. This information density is hardly granular enough to allow the utility manager to track his water drops.

In some cases the meters are themselves missing, electronically. Without sufficient safeguards in place to protect the data, as connections and billing systems grow in the evolution of a utility, it is not uncommon for the electronic records used for billing and the physical installations on the ground to become decoupled. The result is that some connections never make it into the billing system, which means no revenue, increased water loss, and frustration for the utility manager.

Other data errors contributing to NRW and revenue loss include meter parameter errors in the billing system:

- Meter programming inaccuracy (e.g., meter programmed to read in gallons, but entered into the billing record as reading in thousand gallons)
- Incorrect meter size
- Incorrect customer class
- Incorrect ancillary attribution (sewer, garbage, pressure zone, etc.)

Completing a smart grid for water installation allows these discrepancies to be identified and corrected, while providing the systems, processes, checks, and balances necessary to ensure that the physical and electronic systems remain directly coupled.

METER DEGRADATION

Meters certainly degrade over time as a function of component failure, wear and tear, water quality, and so forth. This is an important element of utility NRW control and revenue protection as inaccurate meters directly impact the "sold volumes." Replacing meters with AMI-capable units can have a dramatic impact on utility revenue. In a recently completed smart grid installation, replacing meters resulted in significant increases in the volume of water registered by each meter for billing. In this case, the new meters registered an increase of 6.95 million gallons, a 25.9 percent increase.

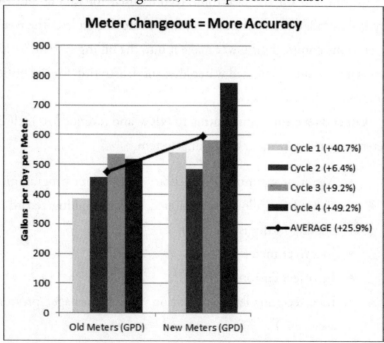

Figure 34: Increases in consumption measured by new meters vs. old meters

But once replaced, understanding the health of the meters becomes critical.

Maintaining a healthy meter population can offer significant revenue protection opportunities for the utility. The question has always been: how does a utility know when a meter is failing?

A smart grid for water installation can answer this question. As the data environment becomes richer, it is possible for meter diagnostics to be performed without actually visiting the meter. That is, we now have the opportunity to use statistical methods to evaluate the integrity of metering systems. A smart grid for water system allows for the first time a means of understanding, in real time, the performance of individual meters within a distribution system.

Borrowing from the study of large-scale networks[119] (ecology, power distribution, etc.), the concept of "critical slowing down" can be applied to metering installations. The result is that we can use this data to proactively identify impending failures of metering systems without knowing the actual condition. And in the process secure utility revenue.

---

119 Scheffer et al., "Early-Warning Signals for Critical Transitions," *Nature 461* (September 3, 2009): 53–59, doi:10.1038/nature08227.

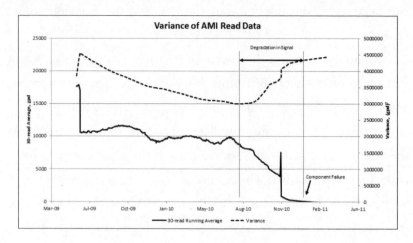

Figure 35: Example variance in read data for a commercial meter

The impact of monitoring for meter degradation is particularly important in large commercial meters. Many people outside the water business do not understand that large meters can represent a significant volume of the water delivered to customers. In one of Global Water's utilities, 10 of the 17,500 meters in the community consume 23 percent of the entire delivered volume. In fact, one meter represents more than 4 percent of the annual consumption. It's clear, then, that ensuring the accuracy, precision, and reliability of these large meters can have a significant impact on revenue.

A recent survey of 50 commercial meters in one utility system found the following:

- Fifty percent of commercial meters tested failed to meet the AWWA meter accuracy standard.[120]
- Sixteen percent of the tested meters registered less than 50 percent of the actual volume passing through them.

---

120 American Water Works Association, Manual 6, "Water Meters: Selection, Installation, Testing and Maintenance"

- Eight percent of the meters in the survey group registered zero volume.
- The annualized revenue loss from the survey group amounted to $280,000.
- The annualized water loss from the survey group amounted to 53,000,000 gallons.

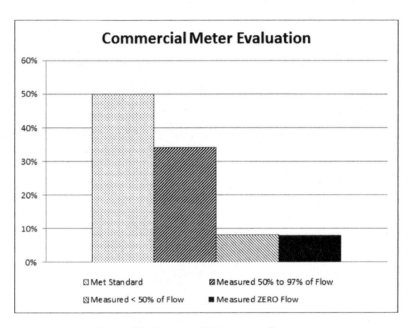

Figure 36: Commercial meter performance

FOUND REVENUE

In the end, the efficiency gains of a smart grid for water installation result in significant water loss reductions, and also provide direct financial benefit to the utility, for example:

- At one utility with 2,500 meters the resulting revenue increase from the upgrade to a smart grid for water metering system was in the order of $11,500 per month.

115

- Monitoring the performance and calibration of a 300-commercial-meter population through smart grid for water systems can protect in the order of $1 million annually.
- Through smart grid for water applications, another utility of 35,000 meters was able to recognize an immediate incremental revenue increase from billing system errors of more than $16,000 per month.

And as the water bill forms the basis for many other charges, the revenue impacts extend beyond to other municipal operations. Throughout the first six months of operation on a smart grid for water installation, one utility realized an annualized benefit of $1.63 million as a result of fixing the data.

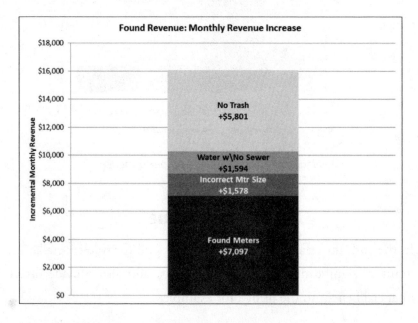

Figure 37: Incremental monthly revenue increases from a smart-grid-for-water installation[121]

121 Global Water, 2012

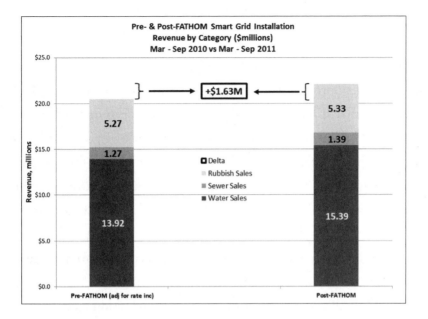

Figure 38: Revenue increases from a smart-grid-for-water installation[122]

In another utility, a smart-grid installation resulted in:[123]

- A decrease in water loss from 34 to 14 percent
- An increase in billed volume of 31.5 percent
- An increase in revenue of 40.6 percent

For the customer, access to highly granular, time-relevant data, means dramatic changes in consumption. With access to this data, subtle societal pressures can be reinforced and the utility can nudge the customers' fundamental understanding of water and their use of it. In the face of increasing water prices, making customers intimately aware of the impact their own actions have on their costs will be imperative.

---

122 Ibid.
123 S. Morris, AWWA ACE Presentation, June 12, 2012

USING THE SMART GRID TO IMPROVE
FINANCIAL PERFORMANCE

Not only does the smart grid for water find revenue, it is also a key system for improving the financial operations of the utility. Many utilities suffer from the related problems of increasing aged receivables and bad debt. In some cases—particularly in poor economic times, or where statutory prohibition of disconnection for non-payment exists—utilities can see accounts become increasingly delinquent, leading ultimately to the write-off of 10 to 20 percent of billed revenue.

While the threat of disconnection provides customers more incentive to remain current on their charges, there are real and significant costs for the utility associated with this action, and real and significant public relations and social issues. As a result, disconnection is viewed as a last resort and many utilities prefer to use softer methods of encouraging payment. In the case of the smart grid for water, by integrating Interactive Voice Response (IVR) technologies with Customer Information Systems (CIS), the payment process becomes much more proactive (eg. reminding customers of their due dates via IVR, and offering payment options such as credit card, direct debit, check etc). The result is that fewer accounts become delinquent, fewer customers are subject to late fees, and customers are afforded flexibility and control over their payments.

This issue is compounded because in reality it is very difficult to collect small bills. And water utility bills are notoriously small. As a function of this and the general lack or inadequacy of systems designed to combat the perils of significantly aged accounts receivable (A/R) or increasing bad debt, the water sector has performed quite poorly in the areas of managing its cash. Consequently, for a sector that is struggling on a cash basis, the sector's A/R metrics are remarkably poor. The reported sector averages are 60-70 percent collected in 30 days, 80-90 percent collected in 60 days with a staggering 5-20 percent written off as uncollectable.

In Global Water's regulated utilities, through FATHOM, aging accounts receivable metrics are impressive: 96.5 percent collected within 30 days and 99.5 percent collected within 60 days with less than 0.5 percent of all revenue written off.

The magnitude of this improvement may not be immediately apparent, but let's consider a simple example of a utility that bills $36 million per year, or $3 million per month. Operating with antiquated systems, the utility collects 60 percent of its revenue in 30 days, and 80 percent in 60 days, and has historically collected only 90 percent of the revenue after many months, resulting in a 10 percent write off.

After implementing smart grid for water technologies like FATHOM, the 30 day A/R metric increases from 60 percent to 96.5 percent and the 60 day A/R metric increases from 80 to 99.5 percent. Write-offs were reduced from 10 to 0.5 percent. In real terms, this means an 11-fold reduction in Days of Sales Outstanding, from 12.4 to 1.1.[124] This has an immediate, significant and permanent effect on cash flow of nearly $2 million. The contemporaneous reduction in bad debt also results in an immediate and permanent improvement in cash, generating $3.12 million in incremental cash in the first year. The combined effect of these two actions generates $5.1 million in incremental cash in the first 12 months of operations.

The impact of increasing the collectable revenue and reducing the time to receive that cash has a remarkable impact on utility operations. In many cases it can significantly improve the utility's financial position, funding multi-year capital improvement plans, or materially reducing debt. In some cases, the combined impact of found

---

124 Days of Sales Outstanding = Accounts Receivable/Credit Sales x Number of Days

revenue and improved cash performance can pay for the investment in smart grid technologies.

USING THE SMART GRID TO PROTECT PUBLIC HEALTH

The smart grid for water can be used to significantly improve the utility's fiscal health, but can also play a major role in protecting the public's health, by providing insight into water quality variations in the distribution system, and dramatically decreasing response times.

By combining customer input via call centers, highly granular consumption data from the customer information system, operational information from supervisory control and data acquisition (SCADA) systems, hydraulic modeling data, and georeferenced spatial asset management data, a rapid, visual identification of water distribution system health can be built allowing operations staff to immediately respond to any potential issue. Further, leak detection flags and reverse flow flags from AMI metering systems can be employed to both identify potential ingress of contaminants, and where hydraulic conditions exist that promote reverse flows.

The smart grid for water also provides immediate and detailed access to information in the case of natural disasters, which can directly safeguard human health and provide for more immediate response and recovery. On May 11, 2011, a magnitude 5.1 earthquake struck Lorca, Spain, causing significant structural damage. The local water utility was able to use the smart grid for water—data integration, electronic metering, and communications—to quickly identify and rectify the impacts on the water distribution system.[125]

- Identification of five major leaks in the distribution system

---

125 Miguel Molina (Agua Ambiente), "Full-Scale Application of Network Monitoring Tools for Leakage Reduction and Asset Rehabilitation Prioritization," presentation to Smart Water Networks Forum, Paris, May 18, 2011

- Identification of twenty-two leaks inside buildings
- Reconfiguration of water distribution system to recover service

Figure 39: Earthquake damage, Lorca, Spain[126]

DECISION SUPPORT SYSTEMS

Using the tools of the smart grid for water, decision support systems aggregating customer water quality/aesthetic issues (CIS) with operational data (SCADA), compliance data (LIMS), flow data (AMI), maintenance data (asset management) and engineering data (hydraulic models) can determine the likelihood, extent, and impact of a potential distribution-system water-quality issue.

The Analytical Water Quality Assurance Program (AWQuA), built from the smart grid for water, allows for early detection and classification of any potential public health issue and proactively

126 © Francisco Bonilla/Reuters, 2011

allows operations staff to identify the necessary rectification plan, while simultaneously allowing compliance staff to notify regulatory agencies. It also allows for automatic customer notification via interactive voice response (IVR) systems, reverse 911 calling, or text messaging. With a rich information system, customer service staff can provide up-to-date information to customers who may call in.

The AWQuA Program also serves as an infrastructure replacement trigger by identifying the most critical areas, from a public health protection and infrastructure reliability perspective.

The most important aspect of the AWQuA Program is the "push" of information to operations, engineering, and compliance staff in order to accelerate investigation and rectification processes.

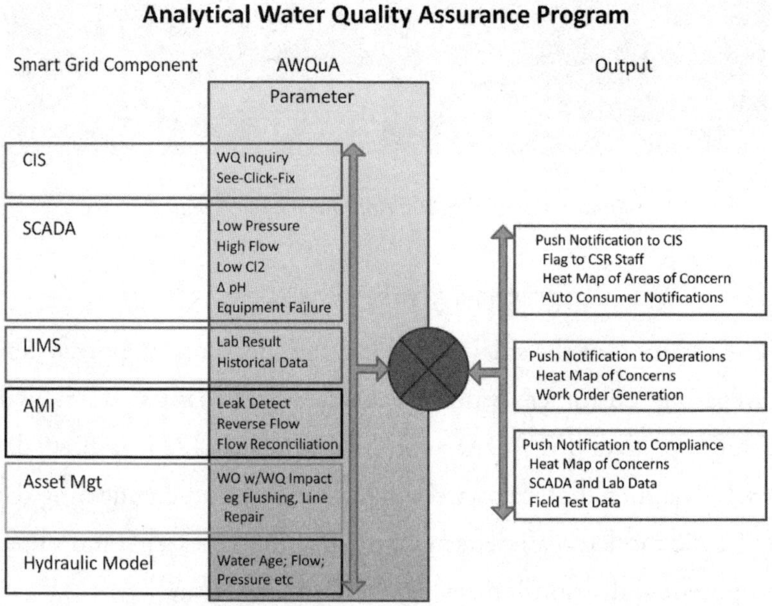

Figure 40: Analytical water quality assurance program

As smart grid for water integrates several major data systems into an information system; it provides immediate relationship information for operators. Each system, customer water quality complaints (CIS), asset management (CMMS), SCADA, AMI, LIMS, and engineering systems (hydraulic models, etc.), provides a specific benefit for monitoring distribution systems.

CIS: Customer Water Quality Complaint Assessments

An often overlooked source for water quality monitoring is the utility's customers:

> Feedback from drinking water customers is particularly valuable to water suppliers, because it is a "real-time" water quality assessment at no cost to the utility. Additionally, these water quality monitors are located at every point in the distribution system where water is being used at all times.[127]

Customers are the first line of defense in water quality monitoring. Unfortunately, their observations are often misdiagnosed or are incorrectly categorized due to the nontechnical nature of both the customer and the customer service representative (CSR) who receives the complaint.[128] In addition, in many cases, the information the CSR has relates only to that call; he or she lacks the geographic and temporal relationships between calls.

---

127 US Army Center for Health Promotion and Preventive Medicine, "Drinking Water Consumer Complaints: Indicators from Distribution System Sentinels," *USACHPPM TG 284*, May 2003

128 The assessment of water quality complaints is an activity requiring specialized knowledge and skill. However, most routinely, the first contact a customer has regarding a complaint is the customer service department. Often CSRs are not located in the same location as the customer or the operations staff, and they typically lack the technical skill to assess the condition.

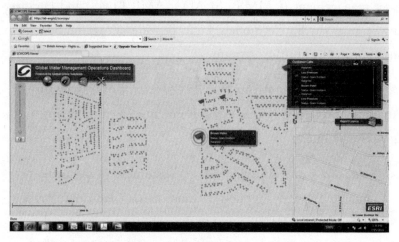

Figure 41: Geographic relationship of water quality complaints

The smart grid for water can provide an instant geographic relationship with other customer complaints, allowing the CSR to provide the customer with up-to-date information regarding any potential problems in the area. By sharing information with asset management and SCADA systems, a more complete response to the customer can be made, and the direct visual representation of problems simplifies and accelerates the response to these events.

Asset Management

In order to address water quality in distribution systems, detailed information on maintenance activities is required. For example:[129]

- Startup or shutdown of treatment processes
- Changes in treatment processes
- Water main breaks
- Water main leaks

---

129 US Army Center for Health Promotion and Preventive Medicine, "Drinking Water Consumer Complaints: Indicators from Distribution System Sentinels," *USACHPPM TG 284*, May 2003

- Sanitary sewer overflows
- Sewer main breaks
- Fire fighting activities
- Distribution system flushing activities
- Storage tank painting
- Construction near waterlines

This information is available to staff through the smart grid for water through links with the asset management system.

AMI Systems

AMI systems not only accurately read meters at a dramatically increased frequency, but also assist in finding and controlling leaks. Understanding the flow dynamics of distribution systems is of critical importance in maintaining public health as they represent an indication of potential contaminant ingress.

Leak detection flags (at customer meters), reverse flow indicators, and inconsistencies between pumped versus billed volumes are all essential information in the assessment of distribution system health.

SCADA/Hydraulic Modeling

Knowing the dynamic regime in which distribution systems operate is critical for understanding the potential water quality implications of events to determine the proper course of action.

Push reports on high flow incidents (fires, etc.), low-pressure incidents (main breaks, etc.) allow operations, customer service, and compliance staff to assess the operational status of the distribution system and to project any water quality impacts.

Laboratory Information Management Systems (LIMS)

LIMS contain current and historical water quality data for both treatment and distribution systems. Access to this data facilitates water quality issue rectification but can also be used to validate and assess past events as models for future events.

SYSTEMS INTEGRATION

Distribution systems lack the comprehensive monitoring programs that are employed on the treatment side of water delivery. That is driven by the fact that while treatment systems can be effectively monitored by one system—SCADA—distribution system monitoring requires the integration of many more information systems, CIS, LIMS, AMI, asset management, and so forth.

Most utilities lack the data and system-level integrations necessary to combine the different types of data (physical, customers, flow, pressure, lab, etc) to make system determinations or to perform system-wide diagnostics. The smart grid for water changes that, and provides the tools to make quantitative determinations on the quality of water in the distribution system, and ensuring public health, while maximizing the efficiency of maintenance and capital expenditure budgets.

USING DATA TO DEFER CAPITAL EXPENDITURES AND REDUCE OPERATING EXPENSES

We have described the "found revenue" aspects of the smart grid for water, but an often overlooked benefit is that the smart grid for water can also "find capacity." This found capacity can result in significant deferral, or even elimination of near- and long-term capital requirements, and can be used to realize significant reductions in operational expenses.

LEVERAGING EXISTING INFRASTRUCTURE

Supply-side management places emphasis on generating new water through increased diversions, massive engineering works or water creation schemes such as desalination. These projects not only require significant investment in time, they are massively expensive. Finding the capital to execute these projects is increasingly difficult for municipalities and customers are becoming increasingly concerned about price increases associated with these projects, the scale of which can be dramatic. For instance, in Melbourne, Australia, residents are expected to see household water bills increase by an average of

34 percent in 2013[130] (or approximately A$300 annually for each household) as a result of the construction of the region's new desalination facility.

Smart grid for water technologies can offer real advantages in this regard. Providing real-time consumption information to customers to encourage conservation, utilities can generate significant water savings, which easily translate into the deferral of construction and water acquisition costs. In Enid, OK, the use of AMI technologies resulted in a reduction of water loss from 34 to 14 percent.[131] In Global's utilities we have seen reductions in demand of 9 to 15 percent.

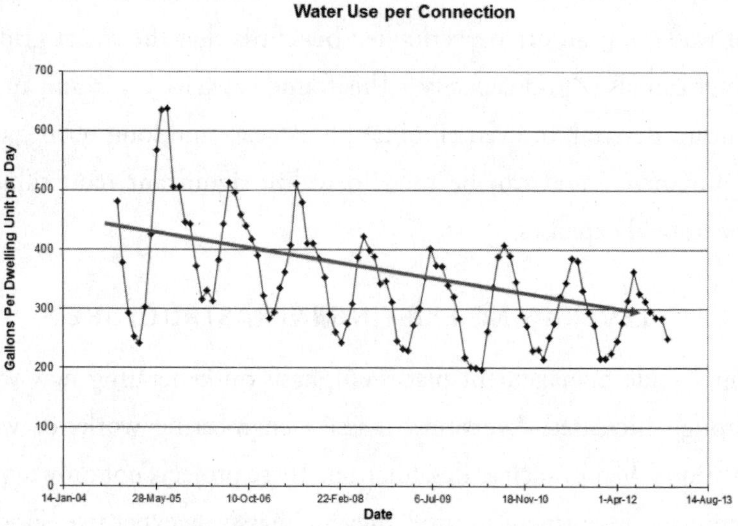

Figure 42: Water demand reduction

These savings leverage existing infrastructure to serve future needs. Because water infrastructure typically scales linearly with

---

130 http://www.abc.net.au/news/2012-11-01/melbourne-households-facing-water-price-hike/4348148, Nov 1, 2012

131 Presentation to AWWA ACE, June 12, 2012

population,[132] these savings can be translated into direct availability for new customers and into significant capital expenditure savings.

For instance, eliminating or reducing leakage through real-time monitoring of consumption, and ensuring all water is billed and water theft is eliminated can provide immediate additional "capacity" for pump stations. For instance, in Arizona, a typical capacity calculation includes average daily flow, a peaking factor assessment, an allowance for leakage and fire flow:

- Average day flow = 250 gallons per unit per day
- Maximum day flow = 495 gallons per unit per day (250 x 1.8 + 10% for potential line losses)
- Peak hour flow = 0.58 gallons per minute (GPM) per unit (1.7 x max day flow)
- Fire flow = 2,100 GPM for four hours

Reducing overall consumption by 20 percent (average daily flow = 200 gallons per unit per day), the maximum day flow would calculate to 396 gallons per unit per day.

As the storage requirements for Arizona are based on the average day maximum month flow,[133] a reduction in demand translates to a reduction in the storage requirements for the existing population, and by extension creates storage capacity in the existing infrastructure for future use.

Similarly, reducing line losses by 20 percent reduces the peak hour flow capacity to 0.57 GPM per unit (a 2 percent reduction). While that may seem small, if we consider the peak hour require-

---

132 L. Bettencourt, J. Lobo, D. Helbing, C. Kühnert, and G. West, "Growth, Innovation, Scaling and the Pace of Life in Cities," *PNAS 104, no. 17* (April 24, 2007).

133 Arizona Administrative Code R18-5-503. Storage Requirements. A. The minimum storage capacity for a CWS, or a noncommunity water system, that serves a residential population or a school shall be equal to the average daily demand during the peak month of the year. Storage capacity may be based on existing consumption and phased as the water system expands.

ments of a 100,000 unit utility, the peak hour savings would be 1,000 GPM less, or in real terms the utility could serve an additional 1,786 units with the same infrastructure.

AGING WATER INFRASTRUCTURE

Water infrastructure in the United States is aging and, coupled with continuing population growth, requires significant investment and stringent focus on cost control. Utilities have no choice but to make infrastructure investment to meet demand and protect public health. Standards, first introduced with the passage of the Clean Water Act in 1972 and the Safe Drinking Water Act of 1974, are becoming increasingly stringent and numerous. For water, the EPA estimates that approximately $334.8 billion in drinking water infrastructure will be needed between 2007 and 2027 in order for water utilities to continue to provide clean and safe drinking water to customers in the United States. For wastewater, the EPA estimates total infrastructure needs of municipally owned wastewater treatment utilities in the United States during the period between 2008 and 2028 at $298.1 billion, of which more than 60 percent represented wastewater treatment, pipe repairs, and new pipes.

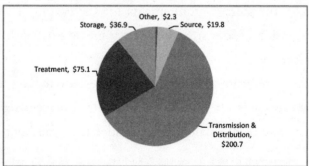

Figure 43a: US water infrastructure needs ($billions)[134]

134 USEPA, "Drinking Water Infrastructure Needs Survey, Second Assessment," Fourth Report to Congress, February 2009

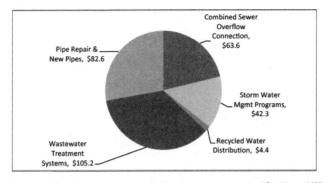

Figure 43b: US wastewater infrastructure needs ($billions)[135]

Unfortunately, access to capital for these activities is becoming increasingly more difficult to obtain, particularly in a period where revenues are decreasing. As a result, it is important to manage these systems as efficiently and effectively as possible to extend the life of infrastructure.

Two methods of achieving this extension are demand shift and pressure management.

BENEFITS OF DEMAND SHIFTING
WITH OFF-PEAK STRATEGIES

By combining highly granular data with time of use incentives, a utility has the ability to spread existing capacity over a larger timescale. This is particularly useful where commercial or industrial users are driving peak demand. By shifting demand, utilities can leverage existing capacity, serving some customers during off-peak hours and freeing existing infrastructure for use during peak hours.

---

135 USEPA, "Clean Watershed Needs Survey 2008," Report to Congress, April 2010

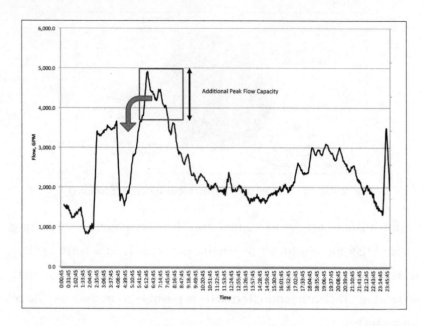

Figure 44: Flow demand shifting

In order to be successful, the utility must be able to demonstrate that the shift makes financial sense to the customer, and can do so by implementing a time-of-use tariff offering incentives to shift. Clearly time-of-use tariffs can only work in the context of understanding the temporal relevance of consumption, and therefore demands that a higher degree of data granularity exist.

PRESSURE MANAGEMENT

Operational CAPEX, those capital investments necessary as a result of operational failures or issues, benefit significantly from smart grid for water installations. Pressure management has become an established leak control methodology for water systems. Reducing these leaks is an important aspect of "finding capacity," but pressure management can also reduce the occurrence of bursts in systems, adding years to the life of existing infrastructure.

Bernoulli's classic pipeline equation demonstrates the demands of flow on pressure:

$$\frac{P_1}{\rho g} + \frac{V_1^2}{2g} + Z_1 = \frac{P_2}{\rho g} + \frac{V_2^2}{2g} + Z_2 + losses$$

By reducing the flow requirements, two benefits are apparent:

- The pressure required to maintain that flow is reduced by a factor of the square of the flow reduction fraction. That is, a 5 percent reduction in flow reduces the pressure requirements by 9.75 percent.
- As line losses are typically directly proportional to velocity, pressure requirement is reduced further.

And by reducing the pressure, we can achieve leak reductions (further reducing the flow requirements) and reductions in main bursts.

LEAK CONTROL VIA PRESSURE MANAGEMENT

Simplistically, leaks from distribution systems can be modeled as the equivalent to an orifice in a pipe or tank. The flow in such cases is largely dependent on the pressure (head) and the orifice coefficient, α. Accordingly a simple relationship can be built that compares the idealized impact of pressure reductions with respect to leak flow reductions:

$$\frac{Q_1}{Q_0} = \left(\frac{H_1}{H_0}\right)^a$$

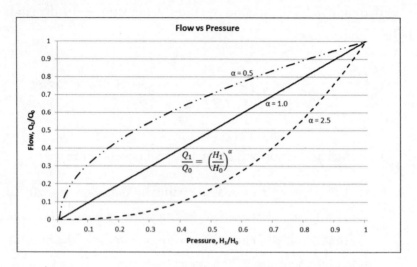

Figure 45: Leak flow reduction via pressure management[136]

In practice, there is a wide variation in orifice conditions, ranging from 0.5 to 2.79[137] and as a result the impact of reducing pressure varies considerably. For instance, reducing the operating pressure of the distribution system from 75 psi to 65 psi (a 13.3 percent reduction) can produce a leak flow reduction ranging from 10 to 30 percent depending on the leak characteristics, laminar/turbulent flow regimes, and the soil/water interface characteristics.

REDUCING MAIN BREAKS

Determining the longevity of pipeline systems in water utilities is a complex task, due to the often unique conditions of the installations, and the various environmental conditions that dynamically exist in the systems. For asbestos-cement pipe, Davis has modeled the failure criterion to be:[138]

---

136 J. van Zyl, and C. Clayton, "The Effect of Pressure on Leakage in Water Distribution Systems", *Proceedings of the Institution of Civil Engineers - Water Management, Vol 60, Issue 2*, 1 June 2007

137 Ibid.

138 P. Davis, D. De Silva, D. Marlow, M. Moglia, S. Gould, and S. Burn, "Failure Prediction and Optimal Scheduling of Replacements in Asbestos Cement Water Pipes," *Journal of Water*

$$\frac{pD}{2b[S_0 - S_R t]} + \left(\frac{wD}{1.048F_m b^2 [S_0 - S_R t]}\right)^2 > 1$$

Where:

p = operating pressure

D = pipe mean diameter

b = pipe wall thickness

w = applied external load

$[S_0 - S_R t]$ = residual strength

F_m = bedding factor

While this model may not be directly applicable to every scenario, it maintains the key elements: age, pressure, wall thickness, internal and external loads. For the utility, there are only two of these variables over which we can exert direct control: residual strength (that is implementing a pipeline replacement program to offset the effect of age and wall thickness deterioration) and pressure.

Obviously replacing pipelines is an enormous and time consuming task—and fantastically expensive. However, by allowing a pressure reduction, we can extend the longevity of our pipeline systems.

ANCILLARY BENEFITS OF OFF-PEAK STRATEGIES

Offering time-of-use tariffs for water customers can have dramatic impacts on the utility's operational power expenses, if the electrical provider is concurrently offering time-of-use opportunities.

Supply: Research and Technology—AQUA 57(4)

Consumptive Charges

As a first order approximation, the power required to sustain a desired flow at a specific pressure can be approximated by:

$$BHP = \frac{Q \cdot h \cdot SG}{3960 \ \eta_m \eta_p}$$

Where:

BHP = Brake Horsepower (foot-pounds/second)

Q = flow (gallons per minute)

h = pressure head (feet)

SG = specific gravity of fluid

η_m = motor efficiency

η_p = pump efficiency

Clearly reducing pressure (h) or flow (Q) can have proportional impacts on the required power. Even small reductions in pressure can result in large power savings. For example, reducing the operating pressure of a system from 75 psi to 65 psi can reduce the distribution system power requirements by 0.065 kWh/DU/day.[139] With an average power cost of $0.09/kWh, the resultant savings are over $200,000 per year in a 100,000 home community.

In addition, if the per capita consumption could be reduced by 20 percent through the provision of conservation messaging and real-time data, the power requirements of integrated water/wastewater/recycled water operations (including power costs for treatment systems, water delivery, wastewater collections, wastewater treatment and recycled water delivery) could result in power savings of $3.2 million over a year for a 100,000 unit municipality.

---

139 Kilowatt-hours per dwelling unit per day

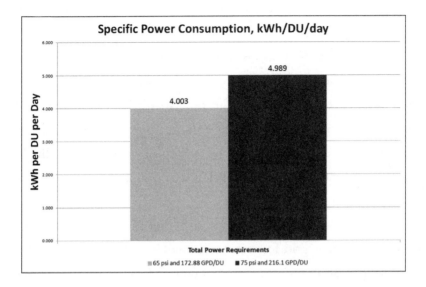

Figure 46: Impact of reducing flow and pressure on power requirements

Where the water utility is itself subject to electrical time-of-use tariffs, shifting the morning diurnal peak to the left or the afternoon diurnal to the right results in power savings of approximately 4.5 percent. Doing both could save this utility 9 percent on power costs.

Demand and Power Factor Charges

We should also note that power charges also include demand and power factor assessments which can be significantly higher than the kWh consumption costs. Power demand charges represent the power utility's cost for maintaining instantaneous capacity availability. In simple terms, a customer pays for the fact that at any time the power utility must be able to meet the customer's instantaneous power demand. Measured as the peak power draw in kW in any fifteen minute window, the costs associated with demand can be very high. Reducing the number of operating pump starts to meet the water utility's demand can significantly reduce these charges. A simple method for achieving this without jeopardizing reliability is to shift water demand (and by extension the operating power) to different times during the day.

Power factor is an assessment made by power utilities to compensate for the real and apparent loads placed on their distribution and generation facilities. Customers with high inductive loads (motors, etc.) can be assessed a significant power factor charge. By reducing flow and pressure, a water utility can reduce the inductive load "seen" by the power distribution network, and therefore reduce these assessments.

From an infrastructure perspective, there are significant opportunities for smart grid for water technologies to expand and extend the use of existing systems. The resultant savings of both operational expenses and capital investment requirements can ensure that our utilities remain fiscally and functionally viable for many years.

FATHOM™ — THE SMART GRID FOR WATER

Global Water has a uniquely developed expert position in the Smart Grid for Water market. FATHOM is a utility-to-utility business that bundles three products, including all hardware, software, implementation services and tax exempt financing to allow municipalities to access state of the art utility management systems.

FATHOM is a utility optimization suite of services that drives efficiencies into utility operations. FATHOM modernizes the data and business processes associated with billing, customer service, asset management and operations by integrating utility data into a geo-temporal data model. Through advanced data analysis and presentment, customers and utility staff are afforded greater insight into consumption use and patterns, allowing sustained resource protection, and increased customer service levels. Global Water uses the FATHOM model to get data to the customer to make the customer aware of their water use, and begin the process of behavioral change.

Through the implementation of FATHOM products and services, utilities save money through operational efficiencies, and they can realize incremental revenue through the identification and rectification of "data leaks".

The FATHOM Advanced Metering Infrastructure (AMI) product is an end to end, integrated solution that provides hourly meter reads and a live customer interface that includes usage alerts, leak notifications, consumption comparisons, and ordinance enforcement. Through the process, a utility is provided with an electronically verifiable meter population that increases data density from one read per month to 720 reads per month, while dramatically reducing the costs associated with read errors and data corruption via advanced read analysis tools.

FATHOM OS (Operating System) allows utilities to access FATHOM's state-of-the-art Customer Information System (CIS) and customer interface systems. This service streamlines customer service, integrates with field work order management systems and is the basis from which customers receive consumption and financial alerts as well as being able to access their data real-time. In addition, FATHOM CIS allows customers access to seven ways to pay their bill.

The FATHOM Asset Management System (AMS) combines real-world spatial data with customer and meter data to allow for streamlined field service operations. In addition, the system can track, schedule and report on all routine and corrective maintenance for utility operations. The AMS is a fully-integrated system that tracks assets in real time to reduce infrastructure failures, and is cloud based, as are all functions of FATHOM.

Each of the FATHOM systems are deployed in a software-as-a-service (SaaS) model, eliminating investments in data systems, IT integration and support and software management.

BENEFITS OF FATHOM

- rapid, instant-on functionality
- dramatically improves the customer/utility interaction
- facilitates reports on utility responsiveness to customers

- supports complex water conservation and individualized rate designs
- provides customer access to highly granular consumption data to encourage conservation
- provides geo-spatial tools to customer service and asset management
- ensures that the meter inventory and characteristics are correct at all times
- provides highly granular meter data for use as a condition monitoring tool
- integrates on-line sensors and/or LIMS data to protect public health
- integrates with sewer provider data

A key benefit of the FATHOM suite of products is the full end to end integration of customer and meter management which provides not only greater efficiency for the utility or municipality, but a much better and coordinated service for the customer.

FATHOM is built on technologies from leading vendors like IBM, NetApp, VMWare, Cisco, Microsoft, Esri and other partners. Virtualization, enterprise storage, blade servers, and Service Oriented Architecture (SOA) allow for quick provisioning and de-provisioning. The environment is hosted in the world's seventh largest data center. The data center is a Tier 3, SSAE No. 16[140] certified environment.

All FATHOM applications are accessible from any location requiring nothing more than a web browser and plug-in.

Where third party applications are utilized in the platform, Citrix has been incorporated to provide a web presentation layer. FATHOM

---

140 American Institute of Certified Public Accountants, Statements on Standards for Attestation Engagements Number 16 (Reporting on Controls at a Service Organization), http://www. aicpa.org/Research/Standards/AuditAttest/DownloadableDocuments/AT-00801.pdf

uses mirroring technologies to replicate data to a secondary vendor storage facility before being backed up to tape. This provides a triple redundancy if a recovery event is required. Customer data protection and security controls are specific for each client. Credentials to access the applications are role based and session logging is incorporated in the event a detailed audit is needed.

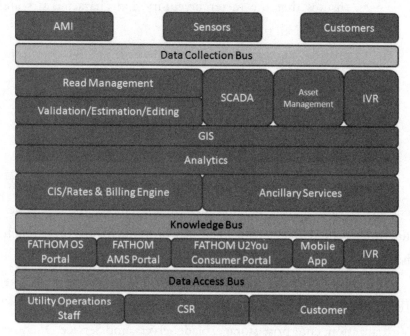

Figure 47: FATHOM Structure

CONCLUSION: A CALL TO ACTION

Together, increased demand and lower supply will place a premium on the industry to find new and more efficient ways of allocating, treating and using water.[141]

-PROFESSOR MARTIN CAVE

For utilities, the convergence of water scarcity and financial pressures will define the twenty-first century. Highly configurable, quickly deployed smart-grid, demand-side management tools are a rapid, flexible, and effective means of reducing pressure on water resources and preserving revenue. At a time when wild variations in natural delivery systems and financial conditions are a constant threat to utility operations, such tools are powerful assets in our arsenal. These systems can be deployed quickly and reliably, resulting in an almost instantaneous demand reduction and the corollary benefits of incremental revenue and life extension for our existing infrastructure. When contrasted against the decades required to permit and construct new water supplies, or the enormous fiscal and environmental costs of water transfer schemes, finding both water and revenue by exploiting our data is a substantially better option.

---

141 M. Cave, "Independent Review of Competition and Innovation in Water Markets: Final Report," London, UK: UK Department for the Environment, Food, & Rural Affairs, April 2009

The benefits of the smart grid for water for nonrevenue water are compelling. But the "found revenue" aspects of the installation can be truly liberating for utilities. In some cases, the installation can be self financing through this increased revenue. The result is a significant improvement in data integrity (better, more accurate, and timelier data), protection and continuity of revenue and overall better customer service.

Adopting a smart grid for water philosophy in our utilities allows for:

- Immediate increase in revenue;
- Immediate improvement in cash flow;
- Immdediate reduction in bad-debt related write-offs;
- Immediate reduction in operating costs:
  - Decreased labor expenses;
  - Decreased costs associated with billing and revenue generation;
- Immediate improvement in customer service;
- Deferral of capital expenditures;
- More efficient water use by customers;
- Real-time understanding of distribution system leakage;
- Rapid identification of anomalous flow conditions (high/low consumption);
- Continual verification of meter health;
- Continual validation of meter and customer inventories;
- More accurate and timely data; and
- Protection of public health.

The smart grid for water also provides the backbone for significant utility optimization programs, including:

- Development of a power optimization plan for water distribution;
- Maximizing efficiency by understanding the real-time peak and average demand;
- Regulatory bylaw enforcement (e.g., water restrictions);
- Allowing for the integration of ancillary devices for increased efficiency:
  - Pressure sensors for outage monitoring;
  - Pressure sensors for power efficiency (e.g., reducing pressure to reduce power);
  - Acoustic leak detectors for leak control;
  - Smart irrigation controllers to better utilize outdoor water use; and
  - Real-time hydraulic flow modeling validation.

Finally, deploying the smart grid for water provides for a significantly improved customer experience:

- Allowing customers to understand their consumption in real time;
- Providing the tools necessary for customers to manage their own consumption and costs;
- Providing push alerts to rapidly curtail leakage and high consumption; and
- Maximizing the benefits of price signals

Maximizing the water management potential of our data systems requires that we reinvent the utility model and focus on engaging the customer and the utility professional in water stewardship. We must fundamentally shift from "supply-side" management, to the much more efficient "demand-side", resource management. As eloquently pointed out by Gary Wolff and Peter Gleick, these are the choices the twenty-first century water utility must make in order to survive:

It is a choice about the path of water development after a basic built water infrastructure has been provided. Should we try to supply more and more water, or is it time to shift our focus from new physical supply to reassessing how fixed supplies of water can be better used to meet ongoing water-related needs? Should water managers stick with the kinds of projects and techniques they know and continue to fail to meet the water-related needs of some people and many ecosystems? Or is it time to emphasize new approaches that seem more likely to meet these needs?[142]

Embracing the benefits of real-time consumption data will take effort. We need to upgrade our metering platforms, inform and educate our customers and regulators, modernize the way we think about information, and ultimately reconnect with the spatial and temporal aspects of water in our lives.

---

142 Gary Wolff and Peter H. Gleick, "The Soft Path for Water," in *The World's Water 2002–2003,* biennial report, Pacific Institute

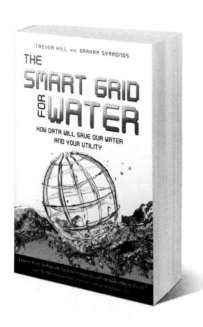

How can you use this book?

MOTIVATE

EDUCATE

THANK

INSPIRE

PROMOTE

CONNECT

Why have a custom version of *The Smart Grid for Water?*

- Build personal bonds with customers, prospects, employees, donors, and key constituencies

- Develop a long-lasting reminder of your event, milestone, or celebration

- Provide a keepsake that inspires change in behavior and change in lives

- Deliver the ultimate "thank you" gift that remains on coffee tables and bookshelves

- Generate the "wow" factor

Books are thoughtful gifts that provide a genuine sentiment that other promotional items cannot express. They promote employee discussions and interaction, reinforce an event's meaning or location, and they make a lasting impression. Use your book to say "Thank You" and show people that you care.

The Smart Grid for Water is available in bulk quantities and in customized versions at special discounts for corporate, institutional, and educational purposes. To learn more please contact our Special Sales team at:

1.866.775.1696 • sales@advantageww.com • www.AdvantageSpecialSales.com